MIX
Papier aus verantwortungsvollen Quellen
Paper from responsible sources
FSC® C105338

Sasmita Panda
Gagan Kumar Panigrahi
Surendra nath Padhi

Earning Animals

Anchor Academic
Publishing

**Panda, Sasmita, Panigrahi, Gagan Kumar, Padhi, Surendra nath: Earning Animals,
Hamburg, Anchor Academic Publishing 2016**

Buch-ISBN: 978-3-96067-080-3
PDF-eBook-ISBN: 978-3-96067-580-8
Druck/Herstellung: Anchor Academic Publishing, Hamburg, 2016
Covermotiv: © pixabay.de

Bibliografische Information der Deutschen Nationalbibliothek:
Die Deutsche Nationalbibliothek verzeichnet diese Publikation in der Deutschen
Nationalbibliografie; detaillierte bibliografische Daten sind im Internet über
http://dnb.d-nb.de abrufbar.

Bibliographical Information of the German National Library:
The German National Library lists this publication in the German National Bibliography.
Detailed bibliographic data can be found at: http://dnb.d-nb.de

All rights reserved. This publication may not be reproduced, stored in a retrieval system
or transmitted, in any form or by any means, electronic, mechanical, photocopying,
recording or otherwise, without the prior permission of the publishers.

Das Werk einschließlich aller seiner Teile ist urheberrechtlich geschützt. Jede Verwertung
außerhalb der Grenzen des Urheberrechtsgesetzes ist ohne Zustimmung des Verlages
unzulässig und strafbar. Dies gilt insbesondere für Vervielfältigungen, Übersetzungen,
Mikroverfilmungen und die Einspeicherung und Bearbeitung in elektronischen Systemen.

Die Wiedergabe von Gebrauchsnamen, Handelsnamen, Warenbezeichnungen usw. in
diesem Werk berechtigt auch ohne besondere Kennzeichnung nicht zu der Annahme,
dass solche Namen im Sinne der Warenzeichen- und Markenschutz-Gesetzgebung als frei
zu betrachten wären und daher von jedermann benutzt werden dürften.

Die Informationen in diesem Werk wurden mit Sorgfalt erarbeitet. Dennoch können
Fehler nicht vollständig ausgeschlossen werden und die Diplomica Verlag GmbH, die
Autoren oder Übersetzer übernehmen keine juristische Verantwortung oder irgendeine
Haftung für evtl. verbliebene fehlerhafte Angaben und deren Folgen.

Alle Rechte vorbehalten

© Anchor Academic Publishing, Imprint der Diplomica Verlag GmbH
Hermannstal 119k, 22119 Hamburg
http://www.diplomica-verlag.de, Hamburg 2016
Printed in Germany

Dedicated to

Our Parents and Teachers

CONTENTS

PREFACE 3

EARTHWORM – VERMICOMPOSTING 4

HONEY BEE – APICULTURE 13

SILK WORM – SERI CULTURE 23

LAC INSECT – LAC CULTURE 35

SHRIMP – SHRIMP CULTURE 55

UNIO – PEARL CULTURE 61

FISH – PISCICULTURE 69

SHEEP – WOOL INDUSTRY 100

PIG – PIGGERY 125

COW – DAIRY 129

AQUAPONICS – AN INTEGRATED FARMING WITH SYMBIOTIC RELATIONSHIP. 139

REFERENCES 152

PREFACE

This book on "Earning Animals" is written with a view to highlight the importance of some invertebrate and vertebrate species used for earning both at individual and national levels.

It aims at creating awareness among students, entrepreneurs and unemployed youth for gainful employment. The candidate species selected are some useful annelids, arthropods, unio, fish and mammals. A chapter on aquaponics- a method of cultivation of fish and plant farming devoid of soil has also been described.

Since most of these animal species are included in the syllabi of Indian Universities and colleges, we believe this book will be helpful to the students to meet their curricular requirements.

While preparing the manuscript, we took the help of many friends and well wishers. To name a few, we are gratefully acknowledge the help and encouragement received from Prof. A. K. Panda; principal, Jatni College, Jatni; prof. U.R. Acharya; Prof. R.C. Choudhury; Dr. T.K. Barik; Prof. S. K. Das; Dr. S Sangeeta and Dr. S. K. Panigrahi. We also thank Dr. A. K. Rath, Dr. N. Panda, Dr. S.R. Mohanty, Dr. R. K. Mohapatra, Smt. Sunita Sarangi and Sri N. Mohanty.

There might be errors due to oversight, we will gladly take care to correct them if pointed out by the readers of the book.

Authors
Sasmita Panda
G. K. Panigrahi
Dr. S.N. Padhi

EARTHWORM –VERMICOMPOSTING

Classification

Phylum: Annelida; Class: Oligochaeta; Order: Opisthopora; Genus: *Eisenia*; Species: *foetida*

Introduction

Deep beneath the earth, they thrive — pink, slimy and insatiably hungry. They are with us all the time, rooting through our gardens, digging through our lawns and consuming everything in their path. Aristotle called them the intestines of the -world. The ancient Chinese called them angels of the soil. Angels or intestines, worms are a tiny but formidable force, eating their way through organic matter and leaving a trail of rich humus in their wake. Vermicomposting is the practice of using worms to turn the organic waste into nutrient-rich fertilizer. In recent years efforts have been made to use the potentiality of earthworms in recycling nutrients, waste management and development of vermicomposting systems at commercial scale. These are also called as "Ecosystem engineers" as they increase the numbers and types of microbes in the soil by creating conditions under which these creatures can thrive and multiply. In India, the integration of crops and livestock and use of manure as fertilizer were traditionally the basis of farming systems. But development of chemical fertilizer industry during the green revolution period created opportunities for low-cost supply of plant nutrients in inorganic forms which lead to rapid displacement of organic manures derived from livestock excreta. The deterioration of soil fertility through loss of nutrients and organic matter, erosion and salinity, and pollution of environment are the negative consequences of modern agricultural practices. In India, millions of tons of livestock excreta are produced annually. Odour and pollution problems are of concern. Currently the fertilizer values of animal dung are not being fully utilized resulting in loss of potential nutrients returning to agricultural systems. The potential benefits of vermicomposting of livestock excreta, municipal solid wastes such as kitchen wastes, market wastes, garden wastes, include control of pollution and production of a value- added product. Vermicomposting of different livestock excreta including cattle dung; horse waste; pig waste; goat waste; sheep waste; turkey waste and poultry droppings has been reported.

Organic wastes can be ingested by earthworms and egested as a peat-like material termed "vermicompost". Recycling of wastes through vermicomposting reduces the problem of non-utilization of livestock excreta. During vermicomposting, the important plant nutrients such as

N, P, K and Ca present in the organic waste are released and converted into forms that are more soluble and available to plants.

Potential benefits of Vermicomposting

Vermicompost appears to be generally superior to conventionally produced compost in a number of important ways;

- Vermicompost is superior to most composts as an inoculant in the production of compost.
- Worms have a number of other possible uses on farms, including value as a high quality animal feeds. Vermicompost also contains biologically active substances such as plant growth regulators. Moreover, the worms themselves provide a protein source for animal feed.
- Vermicomposting and vermiculture offer potential to organic farmers as sources of supplemental income.

Vermicompost has the following advantages over chemical fertilizers.

- It restores microbial population which includes nitrogen fixers, phosphate solubilizers etc.
- Provides major and micro- nutrients to the plants. Improves soil texture and water holding capacity of the soil.
- Provides good aeration to soil, thereby improving root growth and proliferation of beneficial soil microorganisms.
- Decreases the use of pesticides for controlling plant pathogens. Improves structural stability of the soil, thereby preventing soil erosion.
- Enhances the quality of grains/ fruits due to increased sugar.
- Reduces heavy metal pollution by decreasing the metal content in municipal solid wastes (as earthworms absorb all toxic materials like heavy metals such as Hg, Pb, Zn, Cd in their body tissue by vermicomposting).
- At the same time, the beginning of vermicomposting process is a more complicated process than traditional composting:
- It can be quicker, but to make it so generally requires more labour.
- It requires more space because worms are surface feeders and won't operate in material more than a meter in depth.
- It is more vulnerable to environmental pressures, such as temperature, freezing conditions and drought.
- Vermicomposting Technology for Recycling of Organic Wastes.

Methods

In general, there are two methods of vermicomposting under field conditions.
1. Vermicomposting of wastes in field pits.
2. Vermicomposting of wastes on ground heaps

Vermicomposting of Wastes in Field Pits

It is preferable to go for optimum sized ground pits of 20 feet length 3 feet width 2 feet deep for effective vermicomposting bed. Series of such beds are to be prepared at one place.

Vermicomposting of wastes on Ground Heaps

Instead of open pits, vermicomposting can be taken up in ground heaps. Dome shaped beds (with organic wastes) are prepared and vermicomposting is taken up. Optimum size of ground heaps may be 10 feet length x 3 feet width x 2 feet high.

Materials Required for Vermicomposting

- Kitchen wastes, MSW (Municipal solid wastes such as market wastes, hotel wastes), garden wastes, farm wastes etc.
- Fresh cow dung.
- Wastes: dung ratio (1:1 on dry weight basis).
- Earthworm: 1000-1200 adult worms (about 1 kg per quintal of waste material).
- Water: 3-5 liters in every week per heap or pit.

Vermicomopost Preparation under Tree shade by Pit and Heap Methods

Open permanent pits of 10 feet length 3 feet width 2 feet deep were constructed under the tree shade, which was about 2 feet above ground to avoid entry of rainwater into the pits. Brick walls were constructed above the pit floor and perforated into 10 cm diameter 5-6 holes in the pit wall for aeration. The holes in the wall were blocked with nylon screen (100 mesh) so that earthworms may not escape from the pits. Partially decomposed dung (dung about 2 month old) was spread on the bottom of the pits to a thickness of about 3-4cm. This was followed by addition of layer of litter/ residue and dung in the ratio of 1:1 (w/w). A second layer of dung was then applied followed by another layer of litter/crop residue in the same ratio up to a height of 2 feet. Two species of epigeic earthworms viz., *Eisenia foetida* and *Perionyx excavatus* were inoculated in the pit. Moisture content was maintained at 60-70% throughout the decomposition period. Jute bags (gunny bags) were spread uniformly on the surface of the materials to facilitate maintenance of suitable moisture regime and temperature conditions.

Watering by sprinkler was often done. The materials were allowed to decompose for 15-20 days to stabilize the temperature to reach the mesophilic stage, the process has to pass the thermophilic stage, which comes in about 3 weeks. Earthworms were inoculated in the pit or heap with 10 adult earthworms per kg of waste material and a total of 500 worms were added to each pit or heap. The materials were allowed to decompose for 110 days. The forest litter was decomposed much earlier (75 to 85 days) than farm residue (110-115 days). In the heap method the waste materials and partially decomposed dung (1:1 w/w) are made in heaps of dimension; 10 feet length x 3 feet width x 2 feet high and during inoculation channels are made by hand and earthworm @ 1 kg per quintal of waste are inoculated and then watering is done by sprinkler method. Jute cloth pieces are used as covering material.

Suitable species for vermicomposting

There are different species of earthworms viz. *Eisenia foetida* (Red earthworm), *Eudrilus eugeniae* (night crawler), *Perionyx excavatus* etc.

Red earthworm is preferred because of its high multiplication rate and thereby converts the organic matter into vermicompost within 45-50 days. Since it is a surface feeder it converts organic materials into vermicompost from top.

Desirable attributes of worms suitable for vermicomposting
1. Worm should exhibit high biomass consumption together with a high efficiency of conversion of ingested biomass to body proteins, a physiological trait required for achieving high growth rate.
2. Worm should have wider range of tolerance to environmental factors including adaptation to feed on a variety of organic residues.
3. Worm should produce large numbers of cocoons with short hatching time enabling rapid population growth and, linked to this rapid growth, faster composting of organic residues.
4. Life cycle of the worm should be such that mature/ adult phase is quickly reached.
5. Using a mixture of species is likely to be more useful than use of single species.
6. Worm should be disease resistant.

Vermicomposting process: It is an aerobic, bio-oxidation, non-thermophilic process of organic waste decomposition that depends upon earthworms to fragment, mix and promote microbial activity.

The basic requirements during the process of vermicomposting are
- Suitable bedding
- Food source
- Adequate moisture
- Adequate aeration
- Suitable temperature
- Suitable pH

Bedding: Bedding is any material that provides a relatively stable habitat to worms. For good vermicomposting, this habitat should satisfy the following criteria:
- High absorbency: As worms breathe through skin, the bedding must be able to absorb and retain adequate water
- Good bulking potential: The bulking potential of the material should be such that worms get oxygen properly.
- Low nitrogen content (high Carbon: Nitrogen ratio): Although worms consume their bedding as it breaks down, it is very important that this be a slow process. High protein/nitrogen levels can result in rapid degradation and associated heating may be fatal to worms.

Food Source: Regular input of feed materials for the earthworms is most essential step in the vermicomposting process. Earthworms can use a wide variety of organic materials as food but do exhibit food preferences. In adverse conditions, earthworms can extract sufficient nourishment from soil to survive. However earthworms feed mainly on dead and decaying organic waste and on free living soil microflora and fauna. Under ideal conditions, worms can consume amount of food higher than their body weights, the general rule-of-thumb is consumption of food weighing half of their body weight per day. Live stock excreta, viz., goat manure, cattle dung or pig manure are the most commonly used worm feed stock as these materials have higher nitrogen content. When the material with higher carbon content is used with C: N ratio exceeding 40: 1, it is advisable to add nitrogen supplements to ensure effective decomposition. The food should be added only as a limited layer as an excess of the waste many generate heat. From the waste ingested by the worms, 5-10% are being assimilated in their body and the rest are being excreted in the form of vermicast.

Moisture: Perhaps the most important requirement of earthworms is adequate moisture. They require moisture in the range of 60-70%. The feed stock should not be too wet otherwise it may create anaerobic conditions which may be fatal to earthworms.

Aeration: Factors such as high levels of fatty/oily substances in the feed stock or or excessive moisture combined with poor aeration may render anaerobic conditions in vermicomposting system. Worms suffer severe mortality partly because they are deprived of oxygen and partly because of toxic substances (e.g. ammonia) produced under such conditions. This is one of the main reasons for not including meat or other fatty/oily wastes in worm feed stock unless they have been pre-composted to break down the oils and fats.

Temperature: The activity, metabolism, growth, respiration and reproduction of earthworms are greatly influenced by temperature. Most earthworm species used in vermicomposting require moderate temperatures from $10 - 35°$ C. While tolerances and preferences vary from species to species. Earthworms can tolerate cold and moist conditions far better than hot and dry conditions. For *Eisenia foetida* temperatures above $10°C$ (minimum) and preferably $15°C$ are maintained for maximizing vermicomposting efficiency and above $15°C$ (minimum) and preferably $20°C$ for vermiculture. Higher temperatures ($> 35°$ C) may result in high mortality. Worms will redistribute themselves within piles, beds or windrows such that they get favorable temperatures in the bed.

pH: Worms can survive in a pH range of 5 to 9, but a range of 7.5 to 8.0 is considered to be the optimum. In general, the pH of worm beds tends to drop over time due to the fragmentation of organic matter under series of chemical reactions. Thus, if the food sources are alkaline, the effect is a moderating one, tending to neutral or slightly acidic, and if acidic (e.g., coffee grounds, peat moss); pH of the beds can drop well below 7. In such acidic conditions, pests like mites may become abundant. The pH can be adjusted upwards by adding calcium carbonate.

Other Important Parameters: There are a number of other parameters of importance to vermicomposting:

Pre-composting of organic waste: Scientists reported the death of *Eisenia foetida* after 2 weeks in the fresh cattle solids although all other growth parameters such as moisture content, pH, electrical conductivity, C: N ratio, NH_4 and NO_3- contents were suitable for the growth of the earthworms. They attributed the deaths of earthworms to the anaerobic conditions which developed after 2 weeks in fresh cattle solids. It is established that pre-composting of organic waste is very essential to avoid the mortality of worms.

Salt content: Worms are very sensitive to salts, preferring salt contents less than 0.5% in feed.

Urine content: According to Gaddie and Douglas if the manure is from animals raised or fed off in concrete lots, it will contain excessive urine because the urine cannot drain off into the ground. This manure should be leached before use to remove the urine. Excessive urine will build up toxic gases like ammonia in the bedding.

Other toxic components: Different feeds can contain a wide variety of potentially toxic components.
- Detergent cleansers industrial chemicals, pesticides: These can often be found in feeds such as sewage or septic sludge, paper-mill sludge, or some food processing wastes.
- Tannins: Some trees, such as cedar and fir, have high levels of these naturally occurring substances. They can harm worms and even drive them away from the beds. It has been pointed out that pre-composting of wastes can reduce or even eliminate most of these threats. However, pre-composting also reduces the nutrient value of the feed.

Pests and Diseases: Moles prey on earthworms and hence are often a problem when using windrows or other open-air vermicomposting systems. Damage due to rats and moles can be prevented by putting some form of barrier, such as wire mesh, paving, or a good layer of clay, under the windrow. Putting some type of windrow cover (e.g., old gunny bags) over the material will eliminate damage to worms by birds, apart from improving moisture retention and excessive leaching likely during high rainfall events. Centipedes eat compost worms and their cocoons. Fortunately, they do not seem to multiply to a great extent within worm beds or windrows. If they do become a problem, one method suggested for reducing their numbers is to heavily wet (but not quite flood) the worm beds. The water forces centipedes and other insect pests (but not the worms) to the surface, where they can be destroyed by means of a hand-held propane torch. Ants are more of a problem because they consume the feed meant for the worms. This problem can be checked by avoiding sweet feeds in the worm beds and maintaining a pH of 7 or slightly higher. White and brown mites compete with worms for food and can thus have some economic impact, but red mites are parasitic on earthworms. They suck blood or body fluid from worms and they can also suck fluid from cocoons. The best prevention for red mites is to make sure that the pH of the bedding is neutral or slightly alkaline. This can be done by keeping the moisture levels below 85% and through the addition of calcium carbonate, as required.

Sour crop or protein poisoning happens when worms are overfed leading to protein build up in the bedding and production of toxic acids and gases due to protein decay. The better option is to maintain proper feed quality and micro environmental conditions which rule out any possibility of sour crop.

Nutrients in Vermicompost

It has been estimated that earthworms add 230 kg N/ ha/ year in grasslands and 165 kg N/ha/year in woodland sites. Earthworms increase the nitrate production by stimulating bacterial activity and through their own decomposition. There are reports that concentrations of exchangeable cations such as Ca, Mg, Na, K, available P and Mo in the worm casts are higher than those in the surrounding soil. Vermicompost can not be described as being nutritionally superior to other organic manures. Instead, it is a unique way of manure production.

Chemical composition of worm cast:

Organic carbon%	9.15 to 17.88
Total Nitrogen %	0.5 to 0.9
Phosphorus %	0.1 to 0.26
Potassium %	0.15 to 0.256
Sodium %	0.055 to 0.3
Calcium & magnesium (Meq/100 g)	22.67 to 47.6
Copper (mg L^{-1})	2.0 to 9.5
Iron (mg L^{-1})	2.0 to 9.3
Zinc (mg L^{-1})	5.7 to 9.3
Sulphur (mg L^{-1})	128.0 to 548.0

As a processing system, the vermicomposting of organic waste is very simple. Worms ingest the waste material – break it up in their rudimentary gizzards – consume the digestible/putrefiable portion, and then excrete a stable, humus-like material that can be immediately marketed and has a variety of documented benefits to the consumer. Vermitechnology can be a promising technique that has shown its potential in certain challenging areas like augmentation of food production, waste recycling, management of solid wastes etc. There is no doubt that in India, where on side pollution is increasing due to accumulation of organic wastes and on the other side there is shortage of organic manure, which could increase the fertility and productivity of the land and produce nutritive and safe food. So the scope for vermicomposting is enormous.

HONEY BEE – APICULTURE

Classification

Phylum: Arthropoda; Class: Insecta; Order: Hymenoptera; Family: Apidae; Genus: *Apis*; Species: *mellifera*

Introduction

Maintenance of honey bee colonies commonly in hives is known as apiculture. A bee keeper known as apiarist, keeps bees in order to collect honey and other products of the hive which includes **bee-wax**, **propolis**, pollen, **royal jelly** to pollinate crops and to produce bees for sale to other bee keepers. The location where bees are kept is called an **apiry** or bee-yard.

The term Apiculture is derived from the generic name of western honey bee or European honey bee (*Apis mellifera*). The genus *Apis* is a Latin word for "bee" and *mellifera* from Latin *melli*-"honey" and *ferre* "to bear", hence, the scientific name means "honey-bearing bee". This name was recoined as *Apis mellifica* (honey-making bee) by Carlous Linnaeus, after realizing the difference as bees do not bear honey. Bee keeping is an agro based enterprise, which farmers can take up for additional income generation.

Advantages of bee keeping as an income generation activity
- Bee keeping requires less time, money and infrastructure investments
- Honey and bee wax can be produced from an area of little agricultural value
- The Honey bee does not compete for resources with any other agricultural enterprise.
- Bee keeping has positive ecological consequences. Bees play an important role in the pollination of many flowering plants, thus increasing the yield of certain crops such as sunflower and various fruits.

- Honey is a delicious and highly nutritious food. By the traditional method of honey hunting many wild colonies of bees are destroyed. This can be prevented by raising bees in boxes and producing honey at home.
- Bee keeping can be initiated by individuals or groups
- The market potential for honey and wax is high

Products of Apiculture

Honey

Honey is used in cooking, baking, to spread on bread and as an additive to various beverages, such as Tea. Honey is the main ingredient in the alcohol beverage, which is known as Honey Wine OR Honey Beer. It acts as an antimicrobial agent with potential for treating a variety of ailments, as antibacterial with the properties of lowering water activity by causing osmosis, and for chelation of free ions. Honey appears to be effective in killing drug-resistant biofilms which are implicated in chronic rhino sinusitis. Topical honey has been used successfully in a comprehensive treatment of diabetic ulcers when the patient cannot use topical antibiotics. Honey has also been used for centuries as a treatment for sore throats and coughs as an effective soothing agent. Honey is used in the preparation of Face pack and in other cosmetics.

Chemical Composition of Honey

Honey is rich in Carbohydrates and different types of sugars like Fructose, Glucose and Sucrose. It contains Vitamins of B-series (Riboflavin B2, Niacin B3, Pantothenic acid B5), B6, Folate B9 and Vitamin C and also contains different essential metal ions such as Calcium, Iron, Magnesium, Sodium, Potassium and Zinc. Doesn't contain Fat but contains proteins and fibers. 100 ml of honey generates about 304 kcal of energy.

Synthesis of Honey

Honey is synthesized from nectar by honey bees. The bees transform nectar by a process known as Regurgitation and store in hives. The nectar is processed by digestive enzymes in the honey stomach of bees to ingest and regurgitate until it is partially digested. The honey is made concentrated to evaporate water by fanning of bees. Evaporation of water prevents fermentation and increases sugar concentration.

Other products of Apiculture

Bee wax
Formed by worker bees from 4-7 abdominal segments. It is a tough wax formed from a mixture of several compounds e.g. hydrocarbons, mono-di-tri-esters and free alcohol. Used for polishing leather and wooden materials. Used as a mordant for softening wax. Used in skin care products. Used as Bone wax in surgery. Used in decorative items.

Propolis
A resinous mixture, collected from tree buds and sap flowers. Used as sealant of open spaces of the hive. Showing antifungal and antibacterial activities. Used as an anticancer agent.

Royal Jelly
A honey bee secretion, used in nutrition of larvae. Marketed for medicinal use, such as- In homeopathic medicine. In Grave's disease for its immuno-modulatory activity. Anti-inflammatory, wound healing. Anti-cholesterol activity. Prevents vascularization of tumor.

Pollen
It is collected in the pollen basket and carry it to hive. Used as a protein source during brood rearing. An anti-allergic and anti-tumor agent. Provides energy and improves immunity. Lowers stress level.

Geographical Distribution
Currently, there are 28 subspecies of honey bees which are distributed in Africa, Europe, Asia and America. The honey bee originated in Africa and subsequently spread to Europe in two ancient migrations. All species are cross fertile through reproductive adaptations. The adaptations include the behavior of the bees synchronized with environmental conditions in relation to bloom period of local flora.

Bee colonies consist of 3 categories as:
- A queen bee, the only breeding female in a colony.
- Large number of female worker bees (30-50,000 numbers).
- Large number of male drones (thousands) in a colony.

(i) Queen

Only sexually mature female in the hive and all the female workers and male drones are her offspring. Life span of 3-4 years. Capable of laying eggs ranging from 1500-3000/day. Develop from a normal worker bee in radical growth and metamorphosis. Influences the colony by discharging a variety of pheromones (Queen substances), which suppresses the development of ovaries of workers

(ii) Drones

The largest bee next to queen. Male bees of colony without ovipositors and stingers. Do not forage for nectar and pollen. Solely to fertilize the queen. Role in thermo regulation of the hive. Die immediately after mating. Contain haploid set of chromosomes (haploidy). These are descendants of the queen.

(iii) Workers

All females except the queen. Short life span of about 6 weeks. They perform different works like:

- Cell cleaning
- Feeding older and Young larvae
- Collecting nectar and pollen
- Wax making
- Cell building
- Guards of the hive

Life Cycle of Honey Bee

The queen begins egg laying in mid to late winter, so as to prepare for spring which is triggered by longer day length. The queen usually stays inside the hive except for nuptial flight. Queen collects sperm to fertilize up to 1000 eggs. Egg hatches to a small larva which is fed by nurse (worker) bees. After a week the larva is sealed up in its cell and pupate. After a week emerge as an adult bee. On day 16-20, a worker receives nectar and pollen from old workers and stores it. After 20th day a worker leaves the hive and spends remainder of its life as a forager. The population of a healthy hive contains about 40000-80000 bees. Both workers and queen during their larval stage (first 3 days) are fed with royal jelly. Then workers are switched to a diet of pollen & nectar.

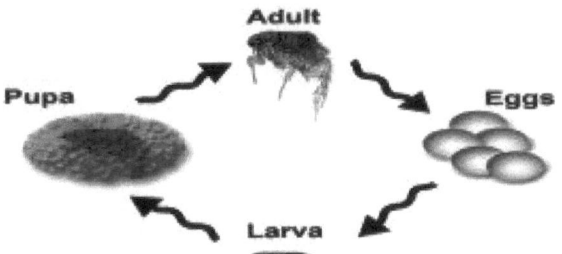

Only the queen continue to receive royal jelly for quick development of larva and pupal stages. Queen is reared in a special chamber of the hive till its emergence. Aged queen bee run for stored sperm, unable to fertilize the eggs due to damaged legs and antenna or else her pheromones have damaged, and can't control all the bees' in the hive. Under such circumstances, the bees will produce one or more queen cells by modifying existing worker cells that contain a normal female egg.

There are two behaviors:
Supersedure queen replacement within one hive without surviving and swarm cell production. The division of the hive into two colonies by surviving.

Supersuder

A highly valued behavioral trait. The hive supersedes the old queen and creates a new queen. The old one fades away or is killed when the new queen emerges. During superseding, the bees produced just one or two queen cells, particularly in the centre of the brood cab. In swarming, a great many queen cells are created, adozen or more and those are located around the edges of a brood cab.

Swarming

The old queen leaves the hive with the hatching of the first queen cell. When she leaves, she is accompanied by a large noumber of bees (young which are wax secretors who can form the new hive secreting wax from the abdominal segments). Sometimes the swarm is accompanied by virgin queens. Often, a number of virgin queens accompanies the swarm and the old queen is replaced as soon as a daughter queen is mated and laying.

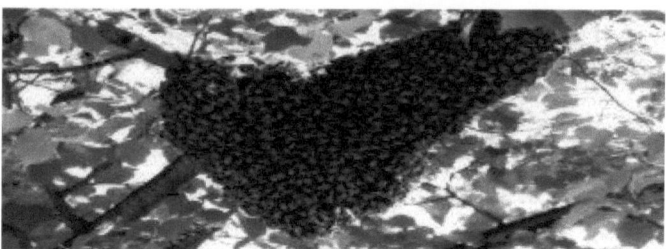

Factors affecting swarming

- Lack of space in the hive (congested nature of the hive).
- Age of the queen (Older queens have swarming pre-disposition).
- Accidental death of the Queen bee.

Honey bee Pheromones

Honey bee's secret special pheromones for almost all behaviors. These pheromones are essential formating, alarm, defense, orientation, and colony reorganization, food production and integration of colony activities.

Honey harvesting

Natural

Collecting honey from wild bee colonies is one of the most ancient human activities and is still practiced in different societies of Africa, Asia, Australia and America. It is as old as 13,000 BC. Honey harvesting from wild hives is done by smoke or by breaking trees or rocks carrying the colony.

Honey bee domestication

Humans began to domesticate wild bees in artificial hives made of logs, wooden boxes, pottery vessels and wooden straw baskets. Honey bees were kept from antiquity in Egypt. In prehistoric Greece, there existed a system of high status apiculture as evidenced from smoking pots, honey extractors and others. Similarly, archaeological findings relating to beekeeping have been discovered in Jordan valley (Israel), Greece, Rome and China.

                          Super frame

Smoking the hive   A capped honey   Uncapping with fork

Extraction & Filtration   Pouring in pots   Packaging

Modern Bee Keeping

In modern times, the Longstroth hive is commonly used. It is a top-open hive with movable frames. These are all square or rectangular structure with wooden frames. It consists of a wooden groove box, floor, crown board and roof. These hives are commonly constructed of cedar, pine or cypress wood. But recently it is replaced by moulded dense polystyrene. Regional variation of hive evolved to reflect climate, floral productivity and the reproductive characteristics of various sub-species of native honey bee in each region.

Country (2005-2008)	Production (1000 metric tons)	Consumption (1000 metric tons)	Number of beekeepers	Number of bee hives
Germany	21.23	89	90,000	1,000,000
Serbia	3 to 5	6.3	30,000	430,000
Denmark	2.5	5	4000	150,000
USA	70.306	158.75	12,029	2,400,000
India	52.23	45	--------	9,800,000
Egypt	16	--------	200,000	2,000,000
New Zealand	9.69	8	2602	313,393
China	299.33	238	--------	7,200,000

Honey production in India and export

About 10,000 tons of forest honey is produced annually. Apiary honey produced under the KVI sector is estimated to be a little less than 10,000 tons in 1990-91. Over 95 per cent of this was from the *A. cerana* colonies, the rest being from the European bee colonies. Forest honey, mostly from rock bee hives, is usually collected by tribals in forests and is procured by forest or tribal corporations as a minor forest produce. Quite a large quantity is also collected by groups or individuals on their own. Forest honey is usually thin, contains large quantity of pollen, bee juices and parts, wax and soil particles. The honey collector gets between Rs. 10 and Rs. 25 per kilogram of the forest honey. Forest honeys are mostly multifloral. Much of the forest honey is sold to the pharmaceutical, confectionery and food industries, where it is processed and used in different formulations. Apiary honey is usually processed at the producers level. This consists mainly of heating the honey and filtering. A few bee keepers or honey producers co-operative societies have better processing facilities that involve killing of honey fermenting yeasts. About 50 per cent of the apiary honey under the KVI sector is graded and marketed under AGMARK specifications. In 1985 the consumption of honey was estimated to be about 8.4 g per capita, while in other countries this was 200 g. Presently this would be about 2.5 g. Honey has so far been consumed mainly as a medicine and for religious purposes. A small quantity has been used in kitchen as an ingredient of pickles, jams and preservatives. With the increasing production in recent years, there is an increasing trend to use honey in food. This is obviously the case with the affluent segments of the population. Forest honey is used in pharmaceutical, food, confectionery, bakery and cosmetic industries.

Sources at the beekeepers co-operative society claim that a beekeeper who invests Rs. 1 lakh for raising colonies (each colony consists of 10,000 to 25,000 worker bees, a queen bee and a few drones) and towards the cost of providing artificial feeding, can realise the entire amount, in addition to profit, within a year. The society has registered moderate sales ranging from Rs 60 lakh to Rs 65 lakh in the past three years. The society finds marketing a Herculean task and has pitched its hopes on the government. A lot depends on the government's move as about 10,000 persons are either directly or indirectly involved in the industry. The government has to set up a research institute to find a cure for the virus that may hit the bee colonies. Apart from beekeeping and marketing, money could be promoted under a self-employment scheme among rural youth in a big way to improve the rural economy, especially when national resources are available in Kanyakumari district of India.

Top Indian Exporters of Honey to the US	
Ranked by Volume January 1 - December 31, 2010	
Company Name	Mtons
1 Nutrisempre Lab De Prod Nat Com.	39,958
2 Eastmate Success Sdn. Bhd.	6,127
3 Kashmir Apiaries Exports	5,213
4 P.T. Burni Panen Raya	4,891
5 Citrofrut S.A De C.V.	4,528

In India bee keeping has been mainly forest based. Several natural plant species provide nectar and pollen to honey bees. Thus, the raw material for production of honey is available free from nature. Bee hives neither demand additional land space nor do they compete with agriculture or animal husbandry for any input. The bee keeper needs only to spare a few hours in a week to look after his bee colonies. Bee keeping is therefore ideally suited to him as a part-time occupation. Bee keeping constitutes a resource of sustainable income generation to the rural and tribal farmers. It provides them valuable nutrition in the form of honey, protein rich pollen and brood. Bee products also constitute important ingredients of folk and traditional medicine.

SILK WORM – SERI CULTURE

Classification

Phylum: Arthropoda; Sub-phylum: Mandibulata; Class: Insecta; Order: Lepidoptera; Genus: *Bombyx*; Species: *mori*

Introduction

The silkworm, *Bombyx mori* is exploited both as a powerful biological model system and also as a tool to convert leaf protein into silk. Sericulture is the breeding and management of silk worms for the commercial production of silk. It encompasses all the activities involved in raising the mulberry plantation, rearing silk worms on the mulberry leaves for obtaining cocoons, reeling the silk thread and twisting it to make it suitable for weaving. It is an underexploited but important sector of the source of income for the beneficiaries with employment generation potential. Further, it also provides employment and income from its ancillary sectors like (i) grainages, chawki rearing centres, seed farms and mulberry nurseries which provide backward linkages and (ii) dyeing, weaving, garment making and marketing. Each such sector can fetch a good return with employment, and human resource management and its utilization. The race of silk worm by which only one crop is taken in one year is called uni-voltine, producing two crops in a year is called bi-voltine and producing more than two crops in a year is called multi-voltine.

History

Historical evidence shows that silk was discovered in China and then this industry spread from there to other parts of the world. The Chinese maintained the secrecy of the beautiful and value added material that they were producing, from the rest of the world for more than 30 centuries. Travellers were searched thoroughly at border crossings and anyone caught trying to smuggle eggs, cocoons or silkworms out of the country were summarily executed. Demand for this exotic fabric eventually created the lucrative trade route now known as the *'SilkRoad'* which was mentioned as early as 300 BC in the days of the Han Dynasty, taking silk westward and bringing gold, silver and wools to the East.

Sericulture in India

India, with a total production of about 15,610 MT that accounts for about 15.13% of global mulberry raw silk production, ranks second among the mulberry silk producing countries of the world next to China (Lakshmi et al., 2011). By the year 2025, domestic demand is expected to increase to 45,000 MT/yr. Therefore, silk production has tremendous growth potential in India, which could provide additional employment opportunities for up to 4 million rural families. Sericulture is cultivated in Karnataka, West Bengal, Tamil Nadu, Andhra Pradesh, Jammu & Kashmir, Gujarat, Kerala, Maharastra, Uttar Pradesh, Rajasthan, Bihar, Odisha, and in other states. Germany is the largest consumer of Indian silk. The muga silk has restricted distribution and is found only in the NE parts of India. Assam is the only state in the country producing all the four varieties of silk. The number of 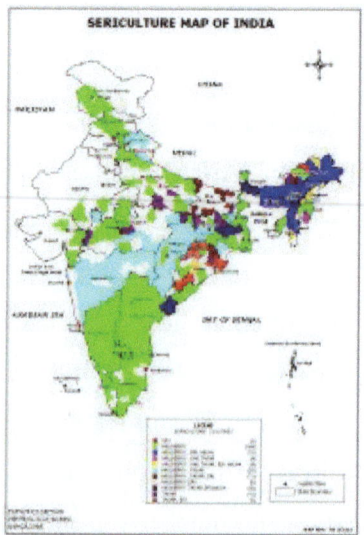 sericulture villages in NE region is about 38,000 and approximately 1.9 lakh families are engaged in this industry in Assam. Similarly, the number of families adopting sericulture had increased to 8,000 with over 10,000 acres of mulberry plantations in Maharashtra in 2008.

Sericulture in Odisha

About 15,000 traditional families involving one lakh people actively practise sericulture in Odisha. It provides indirect employment to equal number of reelers, spinners and weavers. Out of the 4 types of silks viz. Mulberry, Tasar, Eri and Muga cultivated In India, three types namely Mulberry, Tasar and Eri culture is practiced in Odisha. At present with Govt. support tribals and few non-tribals under the BPL category are practicing sericulture and about 50,000 members mostly from SC/ST communities are enrolled under Cooperative fold to earn their livelihood through sericulture. Odisha has rich traditional heritage of tasar culture. The State has the demand of 500MT of raw silk annually for handloom sector with the 10,000 silk handlooms. Present production in the State is about 85 MT. Thus the demand supply gap is wide and emphasis is being given to strengthen the sector through ongoing State Plan Scheme, programme of Central Silk Board etc. Micro project for development of bivoltine mulberry sericulture in Gajapati district in Odisha is under implementation to boost quality bivoltine silk production inside the State of Odisha.

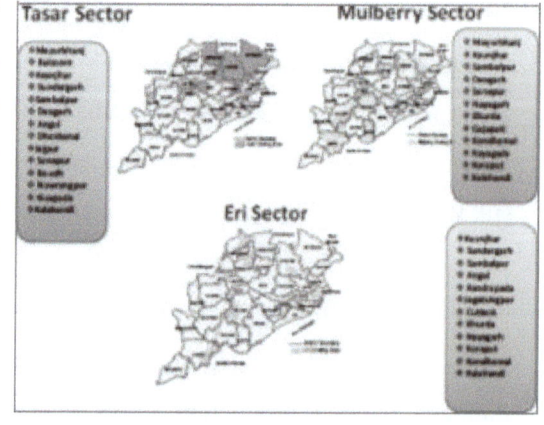

Sericulture activities in Odisha

Life cycle of Silk worm

The silk worm is the larva or the caterpillar of the moth *Bombyx mori* (popularly called as silk moth). The silk worm has four remarkably different developmental stages: egg, larva, pupa, and moth. The total life history of the moth from egg to adult takes about 50 days.

The different stages are as follows:
1. Egg- 10 days
2. Larva (4 Stages) - 30 days
3. Pupa (Cocoon) - 10 days
4. Adult- 2 to 3 days

Adult: The adult silk moth is a creamy white moth that has a flat body and a wing expanse of about 5 cms. It takes no food and seldom attempts to fly. It lives for only 2 to 3 days. After mating, the female moth lays 300-500 eggs on leaves of the mulberry tree.

Eggs: The eggs are round and yellowish-white, and they become grey as hatching time approaches. Two types of eggs are generally found, i.e. diapause type and non-diapause type. The diapause type of eggs are laid by the silk worm inhabiting in temperate regions, whereas silk worms belonging to subtropical regions like India lay non-diapause type of eggs. During diapause all the vital activities of the eggs ceases.

Larvae: The newly hatched larva is about 3 mm long and black in colour. The larvae grow in size and shed their skin (moult) four times. Each growing stage of the caterpillar consumes lot of mulberry leaves. The 5th instar larvae are very important because the larvae have enough nutrients for growth, development and silk production. The last stage full grown larva is about 7 cm long. It has a hump behind the head and a spine-like horn at the tail end. When full grown, the mature larva stops feeding, climbs on a twig and spins a cocoon.

The Cocoon: The cocoon is formed from the secretion of two large silk glands (actually the salivary glands), which extend along the inside of the body and open through a common duct on the lower lip of the mouthparts. The larva moves the head from side to side very rapidly (about 65 times per minute) throwing out the secretion of the silk glands in the form of a thread. The secretion is a clear viscous fluid, which on exposure to the air gets hardened into the fine silk fibre. The cocoons from which moths have emerged are called pierced cocoons. The cocoon consists of only one thread, with a length that can exceed a kilometer. The adult leaves the cocoon by breaking the thread, which is prevented by killing the pupa through heat stifling. The pupa is also killed and dehydrated cocoons are preserved until reeling to draw the thread.

Pupa: The full grown larva pupates inside the cocoon. After about 10 days, it transforms into a winged adult. The adult moth makes an opening in the cocoon and escapes through it.

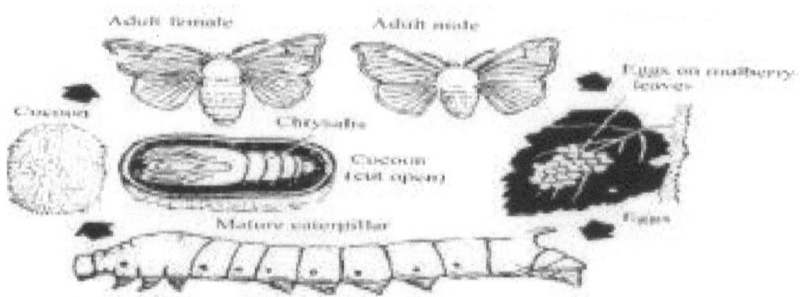

Stages of life history of silk worm

Rearing of silk worms: Selected healthy silk moths are allowed to mate for 4 hours. Female moth is then kept in a dark plastic bed, which can lay about 400 eggs in 24 hours. The eggs are hatched in an incubator. The hatched larvae are kept in trays inside a rearing house at a temperature of about 20°C-25°C. These are first fed on chopped mulberry leaves and after 4-5 days fresh leaves are provided. As the larvae grow, they are transferred to fresh leaves on clean trays, when fully grown they spin cocoons.

Reeling silk: The cocoons are cooked in hot water and the silk fibre is unwound from the cocoons. This process is called reeling. The silk consists of two proteins i.e the inner core is fibroin and an outer cover of sericin. There are four following steps for the completion of the process of reeling:
- The cocoons are first treated by steam or dry heat to kill the insect inside. This is necessary to prevent the destruction of the continuous fibre by the emergence of the moth.
- The cocoons are then soaked in hot water (95° -97°C) for 10-15 minutes to soften the gum that binds the silk threads together. This process is called as cooking.
- The "cooked" cocoons are kept in hot water and the loose ends of the thread are caught by hand.
- Threads from several cocoons are wound together on wheels ("charakhas") to form the reels of raw silk.

Only about one-half of the silk of each cocoon is reelable, the remainder is used as a silk waste and formed into spun silk. Raw silk thus obtained is processed through several treatments to give it the final shape.

Twisting: Prior to weaving, the raw silk is boiled in water to remove remaining gum, dyed and bleached, and then woven into the garment–usually on handloom. In some cases the woven cloth may be dyed and bleached.

Cocoon rearing

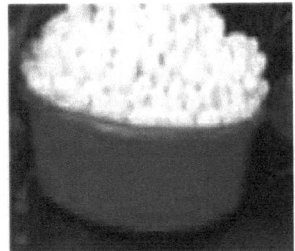
Cocoon harvesting

Species of Silk worms: There are four different species of moths which yield different types of silk:

Mulberry Silk is the most common among them contributing nearly about 95% of world's silk production. It is produced from the cocoons of the moth, *Bombyx mori*. Within the species there are many varieties, mainly differentiated according to the number of generations produced annually under natural conditions. Besides, hybrids of various kinds have also been developed.

Erisilk worm has two varieties – a wild one and a domesticated one bred on castor leaves. The filament is neither continuous nor uniform. Hence the moths are allowed to emerge before commencing reeling. A white or bright red silk is produced.

Tasar silk worms are wild. The Indian Tasar worm feeds on trees of Terminalia species and other minor host plants, while the Japanese and Chinese worms feed on oak and other allied species.

Muga silk worm is found only in Assam. It feeds on two local species of shrubs – *Machilus bombycina* and *Litsae polyantha*, producing a strong, golden yellow thread.

Properties of silk
1. It is bright, soft and strong.
2. It is made of proteins.
3. It is hard wearing.
4. It can be dyed into different colours.

Mulberry cultivation

Mulberry leaves play a very significant role in producing good quality cocoons (Legay 1958). It was observed that better growth and development of silk worm larvae as well as good quality cocoons can be obtained when silk worms fed on nutritionally enriched leaves (Seki and Oshikare 1959). Silk worm obtains 72-86% of their amino acids from mulberry leaves and more than 60% of the absorbed amino acids are used for silk production (Lu and Jiary, 1988). In addition, fecundity of silk worms is also related to larval feeding regime directly and larger pupa produce good quality adults for reproduction which lead to more eggs with high hatchability (Legay, 1958).

The methods of mulberry leaf production vary in different parts of the world depending upon the climatic conditions and soil types. Mulberry is raised as bushes in tropical conditions. The conscious cultivation of mulberry plants for harvesting leaves to be used as food for silk worms is referred to as moriculture.

Mulberry plantation

Climate: The climatic conditions in India are favourable for growth of mulberry and rearing of silk worms throughout the year. For the best mulberry growth, 24 to 26°C temperature and rainfall range of 635 -2500 mm are found suitable.

Varieties: In the past, local varieties of *Morus indica* were grown in the different states like West Bengal, Karnataka and Tamil Nadu for their hardy nature, ability to withstand climatic conditions and quick growing nature. However, their yield was poor. So high yielding varieties have been evolved such as kosen, S 162, S 519, S 523, S 799, C 776, Kanva 2 (M5), S 30 and S 54.

Mysore local and M5 are the two most common varieties used. Mysore local can grow quickly and also withstand climatic variations. M5 is a superior hybrid variety giving higher yield. Further, the leaves are of a better quality with more protein content and give a higher leaf-cocoon ratio. However, it needs supplemental irrigation and fertilizer application.

Preparation of Land and manuring: The field is ploughed deep initially using heavy mould board plough, up to a depth of 12" to 15" in order to loosen the soil. Weeds and gravels are removed. A basal dose of organic manure like compost or farm yard manure is applied @ 10 tonnes per hectare for rainfed crop and 20 tonnes per hectare for irrigated crop. The manure is incorporated by repeated ploughings. This enables easy establishment of the crop. In addition to the organic manure, the inorganic manures or fertilizers used in mulberry cultivation are nitrogen, phosphate and potash in suitable doses as per the requirement of the soil.

Propagation: Mulberry is propagated either through seeds or vegetatively. Vegetative propagation is the most common method of propagation because of various advantages like maintenance of particular characters of the plant. Propagation through seeds has got certain limitations and is used only for breeding new varieties. The exotic varieties which do not come up by cuttings are propagated through root grafts. Many of the indigenous varieties are propagated through cuttings.

Cutting: Cuttings of 7 to 10 cm usually of pencil thickness with three or four active buds are prepared out of the central portion of the clone with the slanting cut. These cuttings are planted in the field directly or in nursery beds. When kept in nursery, all precautions should be observed for not allowing the cuttings to dry up. After 2 to 3 months, sprouted cuttings are transplanted into the field depending upon the type of plantation to be raised.

Grafting: Grafting consists of inserting a small branchof a plant into a rooted plant of the same or allied species in such a way to bring about an organic union between the two and finally make them grow as one. The branch that is inserted is known as scion and the plant to which the scion is inserted is called as stock. The scion grows with the help of nourishment supplied by the stock. The stock is generally of an indigenous variety which is well acclimatized to the local conditions. Grafting thus facilitates the propagation of a varietywhich has the desirable qualities and which cannot be propagated by other means. Exotic mulberry varieties are propagated by this method using root of the local mulberry on stock and shoot of the exotic variety as scion.

Irrigation: The leaf production of mulberry plant is not limited in those areas where there is a uniform rainfall of 100 to 150 mm per month throughout the year. Such conditions does not exist in major sericultural areas, hence supplemental irrigation is essential for optimum leaf yield. Frequency of irrigation varies depending on the growth stages of plant, soil types and other agro-climatic conditions. For example, the frequency varies from once in 8 to 10 days for sandy soils and once in 15 days for clayey soil. More frequent irrigations are necessary for young plants than older ones. The most critical period is during summer when maximum irrigation viz. 10 to 12 days is required.

Pruning: Pruning is the methodical removal of certain branches of a mulberry plant with the object of giving the tree a convenient shape and size, to increase the leaf yield to improve its feeding value. Pruning also helps to divert the energies of the plant for optimum production of foliage. During winter from December to end of February, the mulberry plants remain under dormant conditions. Therefore, mulberry plants are pruned in different ways according to the climate, geographical conditions and forms of the silk worm rearing.

Harvesting of Leaves: The leaves are fed as a whole or as bits and in some cases the entire shoot or branch is used for feeding the worms. Further more, over the past several years the method of harvest must have become modified to suit the availability of labour and intensity of rearing practices. It is recommended to harvest the leaves in the morning hours to avoid active photosynthesis and transpiration during daytime. There are three methods of harvesting the mulberry leaves, namely leaf picking, branch cutting and whole shoot harvest. Usually one or two harvests are done per day and worms are fed four or five times.

Storage and Preservation: A certain time must necessarily lapse during the transport of leaves and after. During this period the leaves are stored properly below 20°C and over 90% relative humidity. The major problem during storage is water loss and breakdown of carbohydrate and protein deterioration in nutritive value. If proper storage at 20°C with 90% relative humidity is achieved, these losses could be prevented.

Diseases of Silk worm

The disease of silk worm may be divided into two classes. Those caused by certain easily recognized animal and plant parasites, not bacteria as " parasitic diseases" and those of more indefinite nature in which bacteria may or may not play a part as "rot disease". Under the parasitic diseases Pebrine, muscardine and fly pest are very important. (1) Pebrine caused by *Nosema bombycis* (2). Muscardine caused by *Beauveria bassiana* and (3) Fly pest caused by *Tricolyga bombycis*. Under rot diseases Flacherie and Grasserie could be the prominent. The parasitic diseases are reasonably well understood and their diagnosis and control are consequently possible. The rot diseases, on the other hand, are some what obscure in their origin and consequently imperfectly understood and difficult to control.

Application of Phytochemical/phytohormone in sericulture

Considering the great bio-diversity of Indian flora, phytochemical research is anticipated to provide potent radiomodifiers and anticancer agents as well as formulations for eco-friendly agriculture and sericulture. A new polysaccharide obtained from the Indian medicinal plant, *Tinospora cordifolia* has been found to possess impressive immunomodulatory and radioprotective properties.

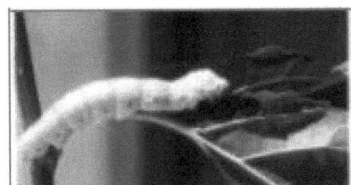

Silkworm feeding on mulberry and phytohormone   *Improved silk cocoons after phytohormone*

A cheap moulting phyto-hormone (MH) preparation that provides better quality of silk in a short time has been developed from an indigenous natural source. The MH formulation was prepared by a continuous extraction technique using an indigenous plant that is widely growing in the coastal region of India. Application of the formulation at a very low concentration (20 ppm) assists in faster and uniform spinning of the silk worms leading to better silk productivity with minimum loss of silk. The product is given to Central Sericulture Research & Training Institute, Mysore for free distribution to silk farmers.

Development of thermo tolerant hybrid strain

The success of sericulture industry depends upon several factors of which the impact of the environmental factors such as biotic and abiotic factors is of vital importance. Among the abiotic factors, temperature plays a major role on growth and productivity of silkworm, as it is a poikilothermic insect. It is also known that the late age silkworms prefer relatively lower temperature than young age and fluctuation of temperature during different stages of larval development was found to be more favourable for growth and development of larvae than constant temperature. There is ample literature stating that good quality cocoons are produced within a temperature range of 22-27°C. Moreover, it is estimated that more than 3000 silkworm strains are available all over the world due to various ongoing breeding programmes Nagaraju, 2002; Thangavelu et al., 2003). Harada (1956) viewed that new silkworm breed has been evolved through hybridization followed by selection. It is important to the farmers to breeding a new good quality silk worm varieties that are thermo-tolerant hybrid strain for rearing. Kumar and Yamamoto (1995) developed a suitable high temperature tolerant biovoltine hybrid i.e HTO5 X HTP5 to cope with high temperature (32±1°C) and low humidity (50±5%) conditions. Now-a-days, large number of thermo-tolerant hybrid strains have been developed.

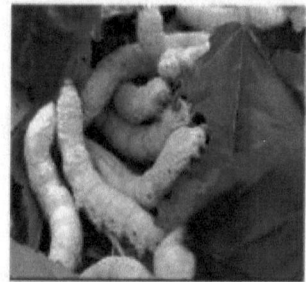
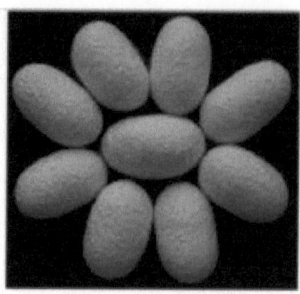

Late age larvae of HTO5 x HTP5   **Cocoons of HTO5 x HTP5**

Application of Nanotechnology in Sericulture

Research and development is now focusing on applications of nanomaterials on human health, energy, defence, catalysis and environment. Efforts are initiated not only towards the health sector but also in the field of agricultural sector (Ulrichs et al.,2006). Surface-modified hydrophobic as well as lipophilic nanosilica could be effectively used as novel drugs for treatment of nuclear polyhedrosis virus (BmNPV) (Barik et al., 2008). Also, research on silk worm, *Bombyx mori* L. race Nistari clearly demonstrates that nano particle could stimulate more production of fibroin protein (Bhattacharyya, 2009).

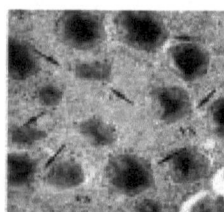
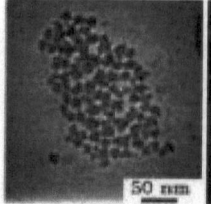

Thin section of polyhedral showing virus particles   **TEM (left) and SEM (right) micrographs of nanosilica (Li et al. 2006)**

Application of Radiation in Sericulture

Irradiation has immense potential and lots of applications (Chanu and Ibotombi, 2011). Gamma irradiation of the eggs of *Bombyx mori* with doses of 2.00-4.00 Gy increased hatchability by 2.4-7.2%, larval survival rate by 3.65-17.93%, mean cocoon weight by 0.42-6.98% and raw cocoon yield by 6.7-16.5% (Petkov et al., 1998). Application of five doses of gamma radiation ranging from 0.01 to 1 Gy at the stage of development of eggs in *Bombyx mori* achieved the greatest effect at a dose of 1 Gy which resulted in a significant increase in larval weight and silk glands by 21.96 and 30.14% resp., an increase in cocoon weight and shell by 11.11 and 9.76%, resp., and an increase in the length of silk filaments and weight by 22.96 and 22.53%, respectively (Abdel-Salam et al., 1995). Irradiation of eggs favourably influenced the fecundity of emerged adults. Eggs (48- and 144-h-old) of *Bombyx mori* when exposed to gamma radiation, the cocoon weight of the 4th generation of insects reared from treated eggs was greater than that of insects from untreated eggs (Rao et al., 1994).

Silk worm Data Base (Silk DB)

The Silk DB is an open-access database for genome biology of the silk worm, *Bombyx mori*, including genome assembly, gene annotation, chromosomal mapping, orthologous relationship and experiment data.

Sericulture of tomorrow

The development and improvement of the current protocol of silkworm transgenesis open new areas of applications both for fundamental research and for applied fields. Transformed silk worms could also be used to study the secretion of foreign fibrous proteins as for example the spider silk with the aim of developing new textile fibres. Silk worms can also be transformed in order to improve sericultural strains. More particularly, it would be very beneficial to produce strains resistant to *baculo*virus infections. This has been initiated by fighting against viral functions through RNA interference and by attempting to increase host tolerance functions against the virus.

Most of the damage to sericulture can be attributed directly to silk-worm diseases, unfavorable weather conditions and poor harvest of mulberry leaves. Therefore, prevention of silk-worm diseases and breeding of a silk worm variety with high productivity are important commercial aspects of sericulture. India with its diverse environmental conditions is known for the local races of silk worms that are rich reservoirs of many resistant genes. These genetic resources can be used for development of disease resistant hybrids in sericulture, and molecular markers as a tool can be used to study the inheritance of such complex traits. Study on the biodiversity of wild silk worm is also needed to protect these genetic resources and their ecologically diverse habitats.

LAC INSECT – LAC CULTURE

Classification

Lac Insect (Lakh ka-kira)

Phylum: Arthopoda; Class: Insecta; Order: Hemiptera; Sub-order: Homoptera; Super-family: Coccidae; Family: Leciferidae; Genus: *Laccifer*; Species: *lacca*

Introduction

Lac is one of the most precious gifts of nature to man and lac insects are exploited for their products of commerce like resin, dye and wax. Since very early times, lac insects and their products have been known to naturalists. As observed by Watt (1908) "Lac enters into the Agricultural, Commercial, Artistic, Manufacturing, Domestic and sacred feelings and enterprises of the people of India to an extent hardly appreciated by the ordinary observers". Today an average of about 20-22 thousand tons of stick lac (raw lac) is produced in the country per year. It pays an important role in contribution of foreign exchange earnings and also provides a subsidiary income to the socioeconomically weakest persons of India.

HISTORY

Since ancient times, Greeks and Romans were familiar with the use of lac. The cultivation of lac insects has a long history in Asia, with some suggestion that it is as old as 4000 years in China where its cultivation accompanied with the development of the silk industry. The lac has been referred in ancient Sanskrit words viz., Atharva-Veda (Dave, 1950; flora, 1952) and was called "Luxa'. The English word lac synonyms Lakh in Hindi which itself is derivative of Sanskrit word Laksh meaning a lakh or hundred thousand. It would appear that Vedic people knew that the lac is obtained from numerous insects. It is mentioned in Mahabharat that 'Luxa Griha' was made up of lac which was prepared by Kaurava for Pandavas. Abul Fazal (1590) in his famous book 'Ain-i-Akbari' has mentioned in detail about the lac industry in India. Mandihassan (1959, 1952), has referred about the lac insect and its products in China. The first scientific reference regarding the lac and lac insect is reported by Kerr and Glover in 1782. Increasing demand of lac products after World War-II has received attention in the present century. In order to increase the production of lac by scientific methods, an association named Indian Lac Association (I.L.A) was formed in 1921, Lac Research Institute (L.R.I) was established at Namkum, Ranchi in 1924, with a view to have greater participation of the Government. In 1930, the Indian Lac Cess Committee (I.L.C.C.) was formed and the committee took over the Indian Lac Research Institute (ILRI) in 1957. Then the need for a

Lac Extension wing was felt and thereafter a Lac Extension Wing (L.E.W.) under the Indian Lac Cess Committee (I..L.C.C.)was created.

DISTRIBUTION

India has its monopoly on the production of lac. Other countries like Africa, Australia, Brazil, Myanmar, Sri Lanka, China, Formosa, France, W. Germany, Japan, Malaya, Nepal, Spain, Thailand, Turkey, U.S.A. and some others also produce lac. But in Thailand, Malaya, Burma and Nepal the lac producing industries are increasing day-by-day. Thailand has become the main competitor of India in export of lac. In India over 90% of lac produced comes from the states Assam (1Cashi Hills), Bengal (Calcutta, Jangipur, Murshidabad, Mathrapur, Malda), Bihar (Manbhum, Palamau, Ranchi, Santhal Pragana), Delhi, Gujarat, Andhra Pradesh , Kashmir, Madhya Pradesh (Damoh, Champa, Bilàspur, Rewa, Umaria), Madras (Coimbatore), Mysore, Odisha (Cuttak, Mayurbhanj), Punjab (Hoshiarpur, Shahpur), Rajasthan (Indergarh, Kota, Jaipur, Thallawar, Karauli), and Uttar Pradesh (Ghazipur, Mirzapur, Agra) etc.

HABITAT AND ITS HOST PLANTS

Lac insects live like a parasite in host plants. It sucks juices from the host plant cells by inserting its beak into the plant tissue. The lac insects have more than one type of host plant. The selection of suitable host plant for the cultivation of lac is much importance. To establish the lac industry one should know well about the topographic and climatic conditions for the growth of host plants suitable for that particular region. Brun (1958) has mentioned that 113 varieties of host plants are found in the geographical Indian regions including Pakistan and Myanmar. Out of these 113 host plants only 15 are very common in India which is as follows:

Table 1: Common host plants of lac insects in India		
1.	Kusum	*Schleichera oleosa, Schleichera trijuga*
2.	Babul	*Acacia nilatica, Acacia Arabica*
3.	Ber	*Zizyphus mauriiianas, Zizyphus jujube*
4.	Palas	*Butea monosperma*
5.	Ghont	*Zizyphus xylopyraKhair*
6.	Khair	*Acacia catechu*
7.	Peepal	*Ficus religiosa, Ficus glomerata*
8.	Gular	*F. glomerata*
9.	Palcapi	*F. virens*
10.	Putkal	*F. globella*
11.	Mango	*Mangifera indica*
12.	Sal	*Shorea robusta*
13.	Shisham	*Dalbergia sisso*
14.	Fig	*Ficus carica*
15.	Arhar or Tur	*Cajanus indicus*

The quality of lac is directly related with the quality of host plant. So far, no artificial product has been able to replace the lac. Khair, Kusum and Babul give better quality of lac when sown directly in the field. But Palas, Ber and Ghont give good crop when they are first sown in nursery and then transplanted to the lac growing fields. Palas and Ber produce a particular type of lac which is called as "KUSUMI LAC".

MORPHOLOGY OF LAC INSECT

Lac insect (*Tachardia lacca* or *Laccifer lacca*) is a minute, resinous, crawling scale-insect which inserts its beak into plant tissues, sucks juices, grow, and secrete lac from the hind end of the body. Its own body ultimately gets covered with lac in the 'CELL". Three types of lac secretory glands are present in lac insects which are on mouth, meso-thoracic spiracle and around the anus. Lac is actually secreted for its protection and not for the food of the insect. The commercial lac is produced in large quantities by female as a protective covering of its body which is injurious to the host plants.

MALE

Male is red in color and 1.2 to 1.5 mm in length. It secretes bright creamy lac. The head region has paired reduced eyes and ten segmented antennae. The mouth-parts are of piercing and sucking type. Thorax bears three pairs of legs and one pair of hyaline wings. The abdomen is eight segmented and terminates into a short, chitinous prominent general sheath containing penis. On either side of this genital sheath a white elongated caudal seta is found. Male insects are enclosed in a cigar shaped thicken capsule having a branchial and tubercular opening at the anterior and posterior end respectively.

Fig. 1: Crimson lac insect: body colour crimson, resin colouryellowish orange.
b, Yellow lac insect: body colour changes to yellow, resin colour remains yellowish orange.
c, Cream lac insect: both body as well as resin colour are creamish.

FEMALE

Female is larger than males and measures about 4 to 5 mm in length. The pyriform body of the female is enclosed in a resinous cell. The head, thorax and abdomen are not clearly distinct. The mouth-parts are of pricking and sucking type. The antennae are clearly visible and degenerated. The posterior end of the body has a median and two lateral processes. The legs are in degenerated form.

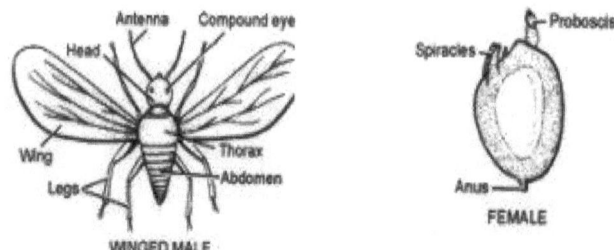

Fig. 2: Structure of male and female lac insect

LIFE HISTORY

The Life cycle of lac insect completed through four stages: egg, nymph instars, pupa and adult in six months.

Fertilization: The reproductions of lac insects have an ovoviviparous type. The females get fixed on the host Plant in resinous mass. The male insect comes out by pushing the operculum of the anal opening, walk over the encrustations of females and fertilize them within their oval cells through anal opening of female. The males leave the parent cell after fertilizing the female. One male is capable to fertilize many females.

Egg-laying and Egg: The fertilization of female is followed by a rapid growth of the female body till it begins to lay eggs. When the time for egg laying is reached, the body of female contracts on the ventral side, gradually vacating a space in the lac cell in which it is enclosed. Lac resin is secreted at a faster rate, and a continuous layer coalesces or grows into one body. After fourteen weeks, the female shrinks in size allowing light to pass into the cell and the space for the eggs. At this time, two yellow spots appear at the rear end of the cell. The spots enlarge and become orange coloured. When this happens, the female has oviposit a large number of eggs in the space called 'Ovisac'. The ovisac appears orange due to crimson fluid called lac dye. It indicates that the eggs will hatch in a week. From these eggs male and female emerge. The male fertilizes the females of this generation and the fertilized female lays eggs and dies secreting lac all the time. Thus the life cycle reoccurs twice in one year on the same host plant. Each mature female after fertilization lays about 200 to 500 small, rounded eggs in a cell in which she is enclosed. The oviposition takes place into the incubating chamber which is formed by the contraction of the body of the female in forward direction inside the lac cell.

Hatching and Nymph instars: After six weeks of laying the eggs are hatched into first instar larvae. At the time of hatching a crimson-red first instar nymph called crawlers come out. The crawler measures about 0.6 x .25 mm in size. When larvae emerge they are in quite large number. This mass emergence of the larvae is known as "swarming" which is continued for 5 weeks. The nymph is 0.5 mm long and 0.25 mm broad, red colored and boat-shaped. The head bears paired antennae, ocelli and ventrally situated piercing and sucking type of mouth-parts. The mouth-parts are provided with proboscis. The three segmented well developed thorax contains two pairs of spiracles and only one pair of walking legs. The abdomen contains two pairs of legs and terminates into a pair of long caudal setae. The active larvae can crawl to a considerable distance just after emergence in search of food and reach their host plants. The young larvae prefer young and succulent shoots as their host plants because they are unable to settle and feed on hard twigs. These larvae settle very close to each other on the twig of the host plant which further collapses completely and form a continuous covering even on the lower surface of the twig. The number of larvae that settle per square inch area is about 150.-200. Settled larvae suck the sap from the twig of the host plant and start to secrete the resinous substance by special dermal glands which are located all over the body. The female larva once settled never moves but undergoes 3 moults inside her cell. After the first moult, both male and female nymphs lose their appendages, eye and become degenerate. While still inside their cells, the nymphs cast off their second and third moult and mature into adult.

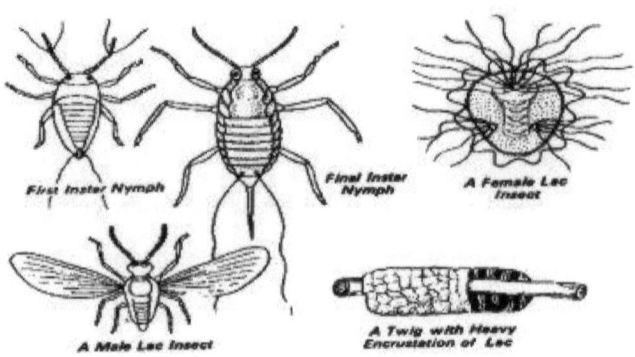

Fig. 3: Life history of lac insect

Pupa: As the resinous secretion comes in contact with air, it soon becomes hard and forms a coating over the body of larva and is called as "cell". Within this cell various life processes like growth of the larva, morphological changes and lac secretion take place. The male 'Cell' is elongated and cigar-shaped having two holes i.e., anterior and posterior. From the posterior hole which is covered by a flap or operculum, the male insect comes out by pushing open the operculum.

Adult: After six to eight weeks of stationary life the larvae are metamorphosed as a result of which some (30%) active winged males and maximum (70%) emerge in the form of females which are wingless. Mostly the proportion of males to females is generally 1:3. Only the male one undergoes a complete metamorphosis; it loses its proboscis and develops antennae, legs and a single pair of wings. The sex is known by the shape of cell even in the early stages of growth. In the case of male, the growth is more on longitudinal axis, whereas in female, the growth is more in vertical axis.

Due to short life period, males do not take major part in the secretion of lac but female secretes lac throughout her life. Life span of female is longer than males. Major quantity of lac is secreted from females. The life cycle reoccurs twice in one year on the same host plant.

The lac insect's life cycle period depends mainly on ecological factors of the region such as temperature, humidity and host plant species.

STRAINS OF LAC INSECTS

Two strains of lac are commonly recognized in India such as Rangeeni and Kusumi. Each healthy Rangeeni female in its life time produces about 0.029 gms and each Kusumi female about 0.069 gm of lac in their life time. The suitable host plants for Rangeeni strains are Palas or Dhak, Ber, Bargad, Peepal, Arhar, Taparia, Siris, Accacia sp. Babul etc. while for Kusumi strains suitable trees are Kusum and Acacia sp. etc. The major crop plants for culture of Rangeeni strain are Kartiki (Oct/Nov) and Baisakhi (May/June) while for Kusumi are Jethwi (June/July) and Aghani (Jan/Feb).

ECOLOGICAL CONDITIONS FOR LAC CULTURE

Ecological conditions play an important role in the cultivation of lac. Srinivasan (1956) summarized three important ecological conditions for its cultivation are:

1. **Availability of plenty of air:** Better lac productions occur in an open area where plenty of air available for hosts plants. But there are also some plants present whose habitat is in conditions where free circulation of air is not possible. So, during selection of new host, it is necessary to choose only those plants that are found occurring in the open or in situations where free circulation of air is secured.
2. **Protection from fire:** Lac cultivation should not be tried in forest fires liable areas.
3. **Host plants:** Long period deciduous species of hot months will not be suitable as host-plants during the summer crop season. They can be taken up for cultivation only for the crops which mature in the cold seasons provided alternative host-plants are available in the vicinity to take on the summer crops.

In addition to the above three important conditions, some other conditions/factor essential for the lac culture are as follows:

(i) Age of Host Plant: Age of the host plants are also another factor that maturation period of host plant varies. The following may be found to be the suitable age for the major host-plants:

Table 2: Suitable age of host plants for lac cultivation

Name of Host Plant	Scientific Name	Age when first suitable for Lac infection
Khair	Acacia catechu	About 8 years
Ber	Zizyphus mauritiana	About 18 years
Ghont	Zizyphus xylopyra	About 18 years
Palas	Butea monosperma	8-10 years
Kusum	Schleichera oleosa	About 15 years

(ii) Climatic condition: For booming lac culture, a warm and moderately dry climate is essential but extreme conditions are always harmful for its production. Three distinct seasons occur i.e. cold, hot and rainy seasons.

(iii) Manuring: Suitable manuring is highly advisable for the healthy establishment of host plants of lac insect which is further not necessary after well establishment. During the absence of artificial manure, natural manure (well rotted cow-dung) may be used.

(iv) Soil and Water: For lac cultivation great care should be taken by giving attention towards soil and water. If soil is found to be deficient in calcium, lime may be added to make up for the deficiency. Water-logging of young seedlings causes mortality, so it should be avoided and at the same time, water conservation for dry month is also advisable.

CULTIVATION OF LAC

Lac cultivation is a complicated process. For its cultivation, proper care of host plants, propagation procession and collection of lac should be taken. So, the cultivators should know well about the following processes:

A. Proper site and host plant selection
B. Pruning
C. Inoculation
D. Swarming, and
E. Harvesting of lac.

A. Proper Site and Host Plant Selection:

The site selection for lac culture should be having more and more plantation of host plants like Babul, Pipal, Palas, Kusum, Khair etc. The selection of suitable host plant for lac insect is of importance, as on these the quality and yield of Lac largely depends. The host plants should be reasonably quick growing, lower sap density, adapted to withstand heavy infestation of the lac insects. The best is 'Kusum' (*Schleichera trijuga*) whose lac commands the highest price. Next to which are 'Pal as' (*Butea frondosa*), 'Ber' (*Zizyphus jujuba*), and 'Sirrus' (*Albizzia lebbak*).

B. Pruning:

It has been seen that the nature and extent of host plants and the time of the year at which pruning is to be done play an important role for bearing and producing feeding grounds for lac insect and lac production. In pruning, proper care of host plant is to be taken so that the host plants don't lose their health, nutrition and produce better quantity and quality of lac. Old and hard branches can never give a satisfactory production. Pruning of trees were done 6 and 12 months before inoculation of lac insect for both Palas and Kusum plant, respectively. Normally, the time of pruning is January-February, for inoculation in June-July, and April-May, for inoculation in October-November. The following points should be taken care of during the pruning:

1. Avoiding of excessive pruning to maintain the general health and strength of the tree.
2. By keeping a good shape of the tree and allowing plenty of room for the growth of new shoots, the cutting of old branches has to be done.
3. Branches exceeding 2" in diameter should not be cut. But cutting at a thickness of 1-2" in diameter will give most satisfactory results.
4. The thin branches under ½" diameter should be cut close to the branches or trunk from which they arise.
5. The dead and diseased branches should be removed and split or broken branches should be cut below the split or break.
6. The lac bearing branches should not be cut indiscriminately but with the aim of serving as the pruning process where necessary.

Types of Pruning:

Two types of pruning have been recommended for lac culture.

   a) Apical/ light pruning: Branches less than 2.5cm diameter are cut from base and branches more than 2.5 cm diameter should be sharply cut leaving a stump of 30-45 cm from the base. This type of pruning is recommended for slow growing conventional tree host species like Palas, Kusum and Ber.

   b) Basal / heavy pruning: Branches having less than 7 cm thicknesses are removed from the base, whereas thicker branches should be cut at a place where it has a diameter of 7 cm. In quick growing bushy host, heavy pruning should be done at a height of 10-15 cm from the ground level e.g. *Flemingia macrophylla, F. semialata*.

Pruning Instruments: The ideal pruning instruments are secateurs and long handled tree prunners. Besides these, pruning knife and Dauli are also used.

Pruning Time Period: Pruning time will need to be adjusted to suit local conditions. In case of Kusum, pruning is best done in the month of June-July and January – February coincide with those in which the crops mature and so harvesting of the mature crop serves the purpose of pruning also.

C. Inoculation:

The first step in the lac cultivation is the inoculation of lac insect. Inoculation is the process by which young ones get associated properly with the host plants. Inoculation is of two types:

1. Natural inoculation: The inoculation taking place in normal routine or in natural way is very simple and common process during which the swarmed larvae infect the same host plant again and start to suck the juices from the twigs. The natural incubation of swarmed larvae has some drawbacks which are as follows:

- *(i) Incomplete nutrition.* Lac insects with their piercing and sucking mouth- parts, pierce into succulent twigs and suck the cell sap of the same host plant for nutrition. If the cell sap of the same host plant is further sucked out by the swarmed larvae of the second crop continuously, the growth of the host plant would be retarded. In this way lac insect may not be able to get enough nutrients from the same host plant. The lac insects due to lack of sufficient nutrients lose their proper development, thereby affecting the production of lac also.
- *(ii) Irregular inoculation.* During the natural inoculation, it is not sure that uniform sequence of inoculation takes place. If inoculation is not of continuous fashion, a regular crop of lac may not be obtained.
- *(iii) Unfavourable climatic conditions.* At the time of swarming a number of factors like high intensity of sunlight, heavy rainfall, flow of wind etc. affect the proper inoculation of larvae. These natural environmental factors may also affect the host plant at the same time and may cause a gap of inoculation resulting in irregularity of the lac crop.
- *(iv) Multiplication of parasites and predators.* Lac insects have certain enemies in the form of parasites and predators. If the crop is not harvested in time and lac is allowed to remain on the same twig, the multiplication of parasites and predators takes place which hampers the population growth of lac insects.
- *(v)* Thus, keeping in view the above drawbacks the natural procedure of inocula-tion is avoided and certain devices have been developed to ensure artificial method of inoculation.

2. Artificial inoculation: The main idea behind the artificial method of inoculation is to check all possible drawbacks of natural inoculation.

In this method, first of all host plant should be pruned in January or June. The twigs bearing insect larvae which are about to swarm, or just before swarming are cut in sizes ranging between 20 to 30 cm in length. Then the cut pieces of these twigs are tied to fresh trees in such a way that each stick touches the tender branch of the tree at several places which form bridges for the migration of the larvae. After swarming, these twigs should be removed and separated from the host plant. The following precautions should be taken in artificial inoculation:

- *(a)* One must ensure that the twigs, which are going to be tied on fresh host plant, are having good number of larvae or eggs. It is also possible that from many of the twigs larvae have swarmed out, thus inoculation would prove unsuccessful.
- *(b)* The twigs provided with eggs or larvae should be without any parasite and predator.
- *(c)* The eggs or larvae present on the twigs should be healthy and about to swarm so that one has not to wait for longer period and thus save time.
- *(d)* For the uniformity of inoculation, 3 to 4 twigs should be utilised.
- *(e)* Host plants should be changed from time to time for the proper nutrition of the larvae.

These insects are very small and if they move to a long distance there are chances of mortality of the larvae. Due to maximum contact of twigs, swarming larvae have not to move for long distance and find suitable places to establish on the host plant.

Inoculation Period: Thus the inoculation periods of all the four types of crops are different. The inoculations of Kartiki, Baisakhi, Agahani and Jethi crops are recommended in months of June to July, October to November, July and January to February respectively. But if continuously four crops are taken, the plant would not get any rest which may cause less production of lac.

Photograph of lac encrusted twig.(Adopted from ILRI Bulletin)

D. Swarming

It is very important phase in the life history of lac insect. So, one should have accurate knowledge about the actual date of the swarming. At the time of swarming, the upper surface has yellow spot on the anal region. At this stage muscle contracts and insect gets detached from the place of attachment. It leaves a hollow cavity which later on gets covered with wax also. When these eggs are to be hatched out they become orange coloured. Thus, it is an indication that swarming has taken place. By trials and learning methods i.e., bypractice one could know about the exact date of swarming by looking at the colour of the eggs.

E. Harvesting of Lac

The process of collection of ready lac from host tree is known as harvesting. In common practice, the harvesting is of two types.

1. **Immature Harvesting:** The harvesting of the lac before swarming is called **immature** type of harvesting and the lac thus obtained is known as 'ARI LAC'.
2. **Mature Harvesting:** The collection of crop after the swarming is called as **mature** harvesting and the lac obtained is known as 'MATURE LAC'.

The harvesting of lac before the swarming has some drawbacks because the lac insects may be damaged at the time of harvesting which would affect the population of lac insects and ultimately result in great economic loss to the cultivators. But in case of Palas lac (Rangini lac), it is found that a lac gives better production. Therefore, Ari lac harvesting is recommended in case of Palas only. In all other cases immature harvesting should be discouraged. It is also found that in cold areas mature crop yields better quality of lac.

The largest yields of lac are obtained by harvesting the infected twigs while female are still living. The harvesting (4 months after innoculation) is done twice a year i.e. June and November. The twig bearing the lac along with eggs is called 'Brood Lac Stick' and the lac is known as 'Brood Lac' or 'Stick Lac'.

Harvesting Period: The harvesting periods of different crops are differ-ent in accordance with the inoculation of crops. Kartiki crop is harvested in October to November whereas, Baisakhi crop in May and June. The other crops like Agahani and Jethi are harvested in January to February and June to July respectively.

RECENT PLAN FOR LAC CULTIVATION

With the increasing number of lac industries some advanced plans have been recommended for the better cultivation of lac crops. Two types of planning are used now-a-days.

1. **Ceupe system:** All the trees of host plants of a definite area are not used under continuous cultivation process of lac crop because if all host plants of a farm would be under continuous attack of lac insects, 100% plants may not get any rest and thus the production of the lac would be affected due to deficiency of nutritive cell sap to the swarmed larvae and adults. So, the plants of a farm are numbered into 5 groups of plants. This artificial division or marking of trees is called ceupe system of crop cultivation. In this system when one group of host plants is under the process of cultivation of lac, other groups of host plants would be under rest.

2. **Alternation of plant:** In this system the variety of host plant is changed after one crop. For example- Khair which is more or less widely distributed is host for both Rangini & Kusumi strai and can be successfully altered with Kusum. So, swarmed larvae are inoculated on the tree of other variety of host plant. In this way every host plant can get enough rest resulting into better production of lac.

PROCESSING OF THE LAC INDUSTRY

When the crop matures fully, most of the lac is harvested and some part is left on the host plant. Forthe proper cultivation, the host plant should be pruned in January every year.

The processing starts with the scraping of the stick lac from the twig. The scraped lac is subjected to removal of many impurities like dead parts of the lac insects, eggs and coloring matter, and finally crushed by hand-operated mortars. Then the material is air dried and obtained in the form of granules which is of pale yellow in color known as **SEED LAC**. This seed lac is soaked in water, washed, dried in sun light, bleached and heated to melt on charcoal fire in cloth bag of 3 to 4 meter. At the time of heating the bag is twisted and the lac is squeezed out of the bag. The impurities of the lac are left out in the bag, and are called **KIRRI LAC**. The squeezed lac is now allowed to cool and solidify around the button- shaped forms which are now called **BUTTON LAC** or **PURE LAC**. This pure lac when stretched into thin sheet is called **SHEET LAC**. This sheet lac when dissolved in water produces white or orange colored lac which is known as **SHELL LAC**. Shell lac is, in fact, prepared by boiling the seed lac with yellow arsenic in a certain proportion. The arsenic is being used for the purpose of improving the colour. The trade grades are: T.N. (The standard grade), fine-orange, garnet-lac, Tongue-lac etc. Lac should not be stored in the green condition as fermentation sets in and fungus growth is reported. Thus the shell lac is most purified form of lac.

The quality of lac depends upon the host plant. Kusumi lac is said to be the best lac while Dhak is supposed to be the worst and cheapest one. The quality and colour of the lac is variable according to the presence of gum and resins in the host plants.

TYPE OF LAC

Lac produced in lac culture is of following types:
- Ari lac - It is immature lac and its use should be avoided.
- Stick lac - It is mature lac harvested in the form of stick is known as stick lac.
- Seed lac - It is obtained after removing and washing from stick is known as seed lac.
- Dust lac - Dust lac is obtained after grinding the seed lac.
- Shel lac - Shel lac is prepared after heating the seed lac and dust lac.

Figure 5: Different types of lac

COMPOSITION OF LAC

Lac is a complex substance having large amount of resins, together with sugar, water and other alkaline substances. The percentages of various constituents are:

Resin	68 to 90%
Dye	2 to 10%
Wax	6%
Albtiminous matter	5 to 10%
Mineral matter	3 to 7% and
Water	3%

The major constituent of stick lac is the resin (70-80%), other constituents like: sugars, proteins, and soluble salts, colouring matter, wax, sand, woody matter, insect bodies and other extraneous matter, a volatile oil is also present in traces.

PROPERTIES OF LAC

- Lac is not soluble in water but easily soluble in alcohol. This property is of great value for insulation of electrical connections.
- Lac is easily fusible on heating.
- Lac has adhesive quality.
- It has binding property when mixed with alcohol.
- Lac is also soluble in weak alkali like ammonia.
- Lac is a bad conductor of heat.
- Capacity of forming uniform durable film.
- Possess high scratch hardness.
- Resistance to water.
- Ability to form good sealers, undercoat primers.
- Capacity to allow quick rubbing with sand paper without slicking or gumming.

ENEMIES OF LAC CULTIVATION

Lac cultivation is destroyed by abiotic and biotic factors:

Abiotic Enemies: These are high intensity of light, high temperature, high humidity, heavy rainfall and flow of wind.

Biotic Enemies: The main biotic enemies of lac cultivation are mammals and insects. Krishnaswami et. al. (1957, 1959), and Gepulpure et.al. (1963), have reported that squirrel, rats, and monkeys cause great damage to the crop. The insects are very powerful enemies of lac crop. The insects damage the crops in different ways.

1. **Parasites:** The lac insects are parasitised by eightspecies of chalcidoid parasites like, *Parenchthrodryinus clavicornis, Erencyrtus dewitzii, Tachardi-aephagus tachardiae, Tachardiaephagus tachardiae var. somervilli, Eupelmus tachardiae, Coccophagus tschirchii, Mariettajavensis* and *Tetrastichus purpureus*.

These parasites lay their eggs into be insects and parasitised 4.8 to 9.9% of lac insects per year and 1/3 of the parasitised cells are males.

2. **Predators:** Predators cause very severe damage to lac cultivation and 35% of the lac cells are damaged by two predators (moth) viz:,*Eublemma amabilis* Moore (Lepidoptera : Noctuidae) and Holocera pulverea Meyr (Lepidoptera : Blastoba-sidae) and three species of Chrysopa (Lacewings) etc. Female lays eggs near encrusted twigs from where larva emerges and feeds on lac insects.

So, these are controlled by proper process of cultivation such as: biological process, artificial process.

Table 3: Species-specific/important insect pests exclusive to lac ecosystem

Insect	Family
Primary parasites	
*Aprostocetus (Syn. Tetrastichus) purpureus (Cam.)	Eulophidae
Coccophagus tschirchii Mahd.	Aphelinidae
Erencyrtus dewitzi Mahd.	Encyrtidae
*Eupelmus tachardiae (How.)	Eupelmidae
*Marietta javensis How. (=M. leoperdina)	Aphelinidae
Parechthrodryinus clavicornis Cam.	Encyrtidae
Tachardiaephagus tachardiae How.	Encyrtidae
T. somervilli Mahd. Encyrtidae	
Predators of lac insect	
Eublemma amabilis Moore	Noctuidae
Pseudohypatopa (=Holcocera) pulverea Meyr.	Blastobasidae
Secondary parasites	
Apanteles tachardiae Cam.	Braconidae
A. fakhrulhajiae Mahd.	Braconidae
Aphrastobracon flavipennis Ashm.	Braconidae
*Brachymeria tachardiae	Chalcidae
*Chelonella cyclopyra Franz.	Braconidae
Elasmus albomaculatus Gahan	Elasmidae
E. claripennis Cam.	Elasmidae
Eurytoma pallidiscapus Cam.	Eurytomidae
Perisierola (=Goniozus) pulveriae Kuerten	Bethylidae
Pristomerus sulci Mahd. & Kolub.	Ichneumonidae
*Trichogrammatoidea nana Zehnt.	Trichogrammatidae
*Not species-specific but with very narrow host range.	

PRECAUTIONS

- Twigs for inoculation should be cut just before the swarming to get healthy brood.
- Twigs used for inoculation should be free from predators and parasites.
- Twigs tied for inoculation should be removed from inoculated host plants after a Maximum period of 20 days.
- Lac left on the host tree for swarming should be removed in October and November.
- The brood be after swarming should be destroyed along with predators and parasites on it.
- The lac scraped from the tree should be taken away from the area of lac infected trees.
- Fumigation and water immersion of lac, before removing from twig, should be done.

Some insect are indirectly beneficial to lac cultivation through keep off the parasites enemy insects. Mostly helpful insects belong to the group of Chaleids, braconoids, ichneumtids and bethylids. These parasitize, the eggs or larvae of *Eublemma anabalis* and *Holocera pulverea*.

YIELD AND COSTS (TRADE)

The annual yield of stick-lac per year per tree may be only 1 to 1.5 kg or as much as 8-16 kg in case of well maintained trees. In the central provinces of India, an average of 340 kg per hectare is said to be oftenobtained from "Palas" Butea trees; taking into consideration of 275 to 290 trees per hectare.

LAC INDUSTRY IN INDIA

India used to produce about 97 per cent of the total lac output in the world but at present it has come down to 50-60 per cent. The production of raw lac in India is approximately 20,000 metric tonnes per year. The cultivation of lac has been a good source as an earner of foreign currency. The major lac-producing states of India are Jharkhand (57% of the country's production), Chhattisgarh (23%), West Bengal (12%), while Odisha, Gujarat, Maharashtra, Uttar Pradesh, Andhra Pradesh and Assam are minor producers. Now-a-days, states like Odisha, Punjab, Madhya Pradesh, West Bengal, Uttar Pradesh, Gujarat, Rajasthan, Assam etc. have shown their increase in production of lac. Over three million tribals inhabiting in these states are engaged in lac cultivation. About 20–38% of the total agricultural income of the tribal growers of Jharkhand is contributed by lac. A lac research institute 'Indian Lac Research Institute' Namkurn, Ranchi had been established in 1925 which is producing good quality of white lac. The Indian white lac is supposed to be better than red or other colored lac because they produce stain or spots at places where they are kept. This is mostly small scale industry with around 350 factories, mostly located in Bihar. Nearly one million man days per annum are generated by the industries engaged in post-harvest processing of lac. Besides, India fetches approximately Rs 120–130 crores of foreign exchange through export of lac every year. Out of total lac produced in India about 85 to 95 per cent is exported specially to Britain, U.S.A., U.S.S.R. and West Germany. Lac resin being natural, biodegradable and non-toxic finds applications in food, textiles and pharmaceutical industries in addition to surface-coating, electrical and other fields and provides immense employment opportunities.

Area Wise Utilization Of Lac Hosts In India Are Given Below:

- Palas *(Butea monosperma)*:It is commonest lac host throughout the greater part of India, extending from North West Himalaya up to 900 m; in hills of South India upto 1200 m.
- Kusum (*Schleichera oleosa*): Throughout Central and South India, Jharkhand, MadhyaPradesh, Odisha and parts of Karnataka and Tamilnadu.
- Ber (*Zizyphus mauritiana*):Important lac host in Murshidabad and Malda districts of West Bengal and Hoshiarpur district of Punjab State, Jharkhand and Chhattisgarh
- Khair (*Acacia catechu*): Only Jharkhand State (Chotanagpur area)
- Ghont (*Zizyphus zylopyra*) :Mainly cultivated in some parts of Northern Madhya Pradesh and Southern Uttar Pradesh
- Jallari (*Shorea talura*): It is the host plant in parts of Mysore and Chennai
- Galwang (*Albizia lucida*) :It is an important lac host in Assam and has also given good results in Chotanagpur area in Jharkhand.
- Ficus spp. (*F. religiosa,F. bengalensis, F. infectoria*): These are universal in occurrence and from which occasionally lac is collected here and there throughout India.
- Arhar (*Cajanus cajan*),Grewia spp. (*G. glabra*and*G.serruleta*), Leea spp. (*L. aspera, L.crispa* and *L. robusta*), Ficus cunia:Favoured host plants in Assam.
- *Moghania macrophylla,Albizzia lucida,Kydia calycina,Ficus rumphii*:Common minor lac host plants of regional importance in Assam.

ECONOMIC IMPORTANCE

The manifest uses of lac is one of the Nature's standing gifts. The various uses to which it is put are:

1. It is of utility to Jewellers and Goldsmiths who use lac as a filling material in the hollows in gold ornaments like bracelets, armlets, bangles and necklaces etc.
2. It is an essential gradient used extensively for making polishes, paints and varnishes for finishing wooden *as* well as metal furniture and doors etc.
3. It is utilized for the preparation of toys, buttons, in pottery and artificial leather.
4. It is used in the manufacture of photographic material, lithographic ink and for stiffening felt and hat materials.
5. It is used as an insulating material for electrical goods.
6. It is mainly used as a resinuous binder in the manufacture of composite moulded insulation materials, and other uses.
7. Moulded insulators like knobs, switch handles, are shields etc., Laminated paper boards and tubes, Laminated sheet and moulded mica, Varnished paper, cloth and silk
8. Insulating and finishing varnished for small coils, Insulating cement or filling compounds, Switch bases and boards, Spark sheets, Use in Surface coatings
9. Spirit Varnish, Aqueous Varnishes, Adhesives and Cements, Sealing Wax, Rubbing Compositions, Gasket Cement, Adhesive for Stainless Steel Tubing, Aniline and Printing Inks, Coating on Earthenware, Wood Turnery, For Glazing Coffee Beans, Under coats and enamels, Electrical goods.
10. It is also used in confectionary trade as anti-fowling compotion for ship bottom, grinding stones and munition fireworks.
11. Lac is also used as a coat for metal ware to prevent tarnishing and for preserving archeological and zoological specimens.
12. Lac is used in process of silvering the back ofmirror.

Lac cultivation not only provides livelihood to millions of lac growers, but also it helps in conserving vast stretches of forests and biodiversity associated with the lac insect complex. Many lac insects and associated fauna have become endangered where lac cultivation has been dumped or habitat destroyed. So, promoting and encouraging lac culture will not only check environmental degradation but also conserve endangered lac insects and associated fauna and flora for posterity. Thus it is of great use and considered to be as one of the cash crops for the cultivators and also to the Government as source of foreign exchange earners which amount to crores of rupees annually. It needs scientific knowledge on lac cultivation so as to increase income and employment generation at the farm level. Improvement in marketing system will certainly be of great help to the lac growers. There is also a need to strengthen the co-operative marketing societies for input as well output marketing.

SHRIMP – SHRIMP CULTURE

Classification

Phylum: Arthropoda; Class: Crustacea; Order: Decapoda; Genus:*Palaemon*; Species: *malcolmsonii*

Introduction

India has vast potential of brackish water area for development of fisheries of fin fish and shellfish along estuaries of 15 major river systems, back waters, mangroves, low lying areas etc. It is estimated that 1.24 million hectares of such potential areas are identified for shrimp culture of which only 16% is being exploited at present. Shrimp culture is mainly dominated by one species *Penaeus monodon* and so it can be termed as mono culture. Unscientific farming with high stocking density and indiscriminate use of drugs and chemicals led to outbreak of diseases like White Spot Syndrome Virus (WSSV),*Monodon Bacculo* Virus (MBV) and bacterial diseases which harmed a lot during late 1990s. There is need to modify the farming activities with guidelines of Coastal Aquaculture Authority to sustainthe farming and earn good profit. It has good potential to generate substantial employment in backward areas (remote coastal villages) with foreward and backward linkages of farming. Different sectors which can generate employment for rural youth and women self help groups are shrimp seed production, farm yard feed production, farm management, harvesting, post harvest care, value addition, transport, net making etc. There is need to enhance knowledge and skill of such people through training for adopting the farming and allied activities. Careful introduction of *Litopenaeus vannamei* and crop rotation with culture of mullet and mud crab are suggested for improving sustainability. Agriculture sector is considered as back bone of Indian economy because two third of people are engaged in it. Share of agriculture GDP in national GDP had declined from 35% in 1980-81 to about 18% in 2010, but proportion of population depending on it decreased insignificantly from 70% to 66%. The decline in profitability in agriculture during the decade 1990s was 14.2%. Hence there is need to modify and diversify farming activities with emphasis on high value farming like horticulture and aquaculture. Innovative farming coupled with value addition and agro-retailing has vast scope for development and thus to generate employment substantially in rural areas. It can also attract unemployed youth and women SHGs to farming by making it intellectually stimulating and economically rewarding. World fish production is 142 million tons with India's contribution of 8.03 million tons (5.33%). India occupies second position in fish production after China in Global level. Its aquaculture production is 3.96 million tons

(80% of inland fish). World per capita fish production was 8 kg in 1950s and 15.8 kg in 1999 (Menon, 2007) and it increased to about 17 kg at present. However, India's per capita fish consumption is 9.9 kg which is below the recommended level of 11.1 kg of WHO. Fish production in developing countries was 60 million tons versus combined production of meat of 70 million tons (beef, sheep, pig, poultry). More than one billion people depend on fish as primary source of protein and supply 15% of animal protein to 3 billion people. Contribution of fish to animal protein in different continents includes 28% in Asia, 21% in Africa, 10% in West Europe, 8% in Latin America and 75% in N. America. In Countries like Bangladesh, Maldives and Pacific Islands, its magnitude is about 755 of animal protein. Nearly 50 million people in world (10 million in India) earn their livelihood from fisheries in 2004 (Menon, 2007).

Aqua-farming and its Importance

Different types of aqua-farming activities having potential to generate employment includes (i) carp culture and composite farming, (ii) scampi culture (iii) catfish culture (iv) murrels culture (v)ornamental fish culture (vi) pearl culture (vii) exotic fish culture (viii) integrated farming (ix)Brackish water fish culture (x)shrimp culture (xi)Mari-culture (xii) molluscan aquaculture (xiii) Enclosure farming with cages andPens (xiv) waste water aquaculture (xv)Aqua-farming for Public health (xvi)cold water/upland aquaculture (xvii)sports fisheries (xviii)export oriented aqua-farming etc. All these activities has several economic and ecological advantages which includes (i) income enhancement (ii) employment generation (iii) production of protein and nutrient rich food (iv)fish for health (v) Recharging underground aquifer (vi) converting detritus and organic wastes in to fish (vii) controlling weed infestation and siltation and thus aging of water bodies (viii) larvicidal fish controlling malaria parasites and help to public health (ix) insect control in paddy fields (x) plants around bundhs promote green environment (xi) improvement of tourism (xii) promotes conservation of aquatic biota (xiii) reduction of local warming in summer (xiv) reducing local pollution (xv) integrated farming providing food and nutrition security to farmers etc. Considering all such aspects, it can be inferred that aquaculture plays a significant role in food security, nutrition security, income security, employment security, health security, water security and ecological security of current society. So it needs attention of all concerned for development in a sustainable manner.

Genesis and Developments in Shrimp Farming

Brackish water shrimp culture is identified as one of farming activities in remote and rural areas of coastal belt of India. It can uplift economic condition of local people b providing them employment. It has long history development from *Bherries* fisheries and *Pokkali* paddy field fisheries of Kerala. In the former case, tidal water is impounded in intertidal mud flats through erected bundhs. Fish and shrimp seeds are trapped through sluice in high tide and allowed to grow with natural feed which yield about 700 kg/ha/year. In the latter case, the low lying paddy fields of Kerala were used for culture of salt resistant paddy *Pokkali* in monsoon months. After harvest of paddy, farmers use it for shrimp and fish culture in post monsoon period by natural trapping of seeds in spring tide. Experimental brackish water fish and shrimp culture started in 1970s in CIFRI and CMFRI centers. After realizing high values of shrimps, its culture started in extensive and modified extensive manner in coastal areas during early 1980s. Large scale development occurred from 1988-89 with establishment of commercial shrimp hatcheries by MPEDA. Attracted by its high profits and short gestation period, farmers and entrepreneurs adopted improved farming practice resembling semi intensive farming in early 1990s. It involved stocking of 1 to 3 lac seeds /ha, regular water exchange and aeration, high quality pellet feeds etc. which resulted production of about 5tonns/ha/5months. There are different types of shrimp culture (Das and Padhi, 2010) based on management practice as follows.

Improved Traditional: It is practised in tide fed ponds in traditional manner with stocking of about 40000 to 60000 seeds/ha/crop. Selective stocking and occasional feeding with local feeds are used to increase production.

Modified Extensive: It can be adopted in either tide fedor pump fed ponds. Stocking density is more than traditional farming and Shrimps are fed with pellet feeds. Water exchange is done.

Semi-Intensive: Seeds are stocked at 1 to 3 lacs /ha/crop. Water quality is monitored with regular water exchange and aeration. Feeding management is practised with high protein pellet feeds. Health of shrimps are monitored regularly.

Intensive: Shrimps are cultured under fully controlled conditions with high stocking density of 5 to 10 lac seeds/ha/crop. Advanced farm management practice like daily feeding with high protein pellet feeds, aeration, water exchange, surveillance of pathogens , water quality monitoring etc. are followed. Such farming is not paractised now due to several adverse effects. India was doing well in farming of tiger shrimp till mid 1990s. Then occurred epidemic dreadful white spot syndrome virus (WSSV) along with *monodon bacculo* virus (MBV) disease and other bacterial diseases causing severe damage to shrimp farming in 1995. It was attributed to indiscriminate stocking, injudicious use of inputs (feed, chemicals, drugs etc.). Then PIL (Public Interest Litigation) was filed by environmentalists citing its damage to environment. So the Apex Court intervened which led to establishment of Aquaculture Development Authority in 1997 under Ministry of Agriculture. Subsequently Coastal Aquaculture Authority of India was established at Chennai to regulate activities of shrimp and fish farming in coastal belt of India. Shrimp production increased from about 35000 tons to little more than 80,000 tonns form 1991 to 1995. Then it exhibited declining phase from 1995 to 2000. Then it started increasing from 2001(approx.1 lac tonns) to 2007(apprx.1.4 lac tonns). Shrimp production exhibited downward trend from 1.4 lac tonns in 2006-07 to 1.06 lac tonns in 2007-08 due to occurrence of shrimp diseases. It is mainly due to sole dependence on one species *Peneaus monodon*.Then it started declining to about 75,000 tons during 2008-09. Farm gate price of *P.monodon* declined drastically in 2007 to Rs200-220/Kg for 25-30g size which was attributed to low priced *L. vannamei* produced from China, Indonesia and Vietnam (Ponniah, 2010) and it affected the sector adversely.

Resource and Productivity in Shrimp Farming

Brackish water resource of India includes 3.9 million ha. of estuaries of 15 major river systems (>20,000 sq. Km. Catchment area), 3.5 miilion ha. back water. Brackish water area suitable for shrimp culture in India is about 1.24 million hectares. Potential suitable area for culture is highest in West Bengal (34%) followed by Gujurat (32%), Andhra Pradesh (13%), Maharastra (7%), Kerela and Tamilnadu (5% each), Odisha (3%, Goa (2%) and Karnataka (1%). But 16% of such vast potential is under culture now of which 4% under traditional manner. Shrimp aqua-farming production increased from 0.8 million tons in 1991 to 3.3 million tonns in 2007. China (40%) tops the list followed by Thailand (15%), Vietnam (12%). India ranks seventh in the list contributing 3% of it. Twenty species of shrimps are cultured in the world with six species dominating the farming sector. They are *Penaeus monodon, Litopenaeus vannamei, Fenneropenaeus indicus, F.merguiensis, F.chinnensis* and *Marsupenaeus japonicus*. It was found that *L.vannamei* (white leg shrimp) is gradually replacing *P. monodon* (tiger shrimp) in several areas. So, tiger shrimp production had declined from 60% to 18% from 2002 to 2007 while that of white leg shrimp increased from

20% to 70% during the same period. SPF (Specific Pathogen Free) stock of the latter accompanied with high production potential are reasons behind it.

Issues in Shrimp Farming & its Diversification

Major issues before shrimp farming are prevalence of white spot syndrome virus (WSSV) disease. It is transmitted from wild stock brood stock to larva. Shrimps do not have specific immune system like fish and vertebrates to respond vaccination. So prevention is advisable through better management practice (BMP) to control the disease. Sea water used in hatchery needs to be disinfected by chlorination and subjected to UV filtration. Disinfection protocol should be followed in the case of implements and personnel working in hatchery. Pathogen free seeds must be used for culture. The BMP usually intended for promoting stress free environment. It includes removal of organic wastes from pond bottom and subjecting it to sun by ploughing for improving mineralization. Water needs to be exchanged at regular intervals and aeration should be practised. Optimal feeding needs to be followed to avoid feed waste and subsequent pollution. Surveillance of pathogens is needed and experts must be consulted immediately in case of out break of any disease. In order to reduce occurrence of diseases in shrimp culture, there is need to shift from monoculture of tiger shrimp to its diversification. Forty species are cultured in China. There is need to culture *Fenneropenaeus indicus, F. merguiensis, Marsupenaeus japonicas* shrimps in culture system intermittently as per suitability. Farmers should also culture seabass *Lates calcarifer* and mud crab *Scylla serrata* because its seed are available in hatchery. Technology should be developed for culture of mullets and pearl spot in such ponds. Now an exotic species with high growth potential and export value is being identified i.e. *Lito penaeusvannamei* which can be cultured following guidelinesof CAA. It is a potential candidate species in USA. Then it was introduced in China, Indonesia, Thailand etc. with good success. It is SPF (specific pathogen free) stock and thus less susceptible to diseases. They can be well farmed under biosecured condition at initial stages which includes net fencing or covering the pond, installing aerators @ 1 HP/0.5 ton biomass, good feeds etc.World Aquaculture Authority (WAA) sponsored " International Meet on Asia Pacific Aquaculture" at Kochi (17-20, January,2011), manyissues (fears) were raised by scientists and aquaculturists on vannamei scenario in India. A brief comparative account of *P. monodon* and *L.vannamei* farming (Felix,2011) is given here. Parameter *P.monodon, L. vannamei* Stocking density/ha 1 lac (10/m) 6lacs (60/m) Survival @ 70% 70,0000 4,20,000 Days of culture (DOC) 150-160 days 110-120 days Yield @35-40g 2.5 to 2.8 tonns/ha/crop @18-20g 7.5 to 8.4 tonns/ha/crop Culture of*L. vanammei* is expanding very fast due to high production and profit. It has grown from about 20% in 2002 to 70% in 2007. So, its culture can be made popularised in India following guidelines of CAA.

Employment Generation

Coastal villages are considered backward due to lack of infrastructural facilities like roads, electricity, schools, hospitals, markets etc. Fisher communities residing in such areas are socially, economically and educationally backward. There lies huge magnitude of low lying, swampy and marshy lands which are not suitable for agriculture. So ,it is lying idle in most of areas. Shrimp culture is a suitable option for utilization of such unutilized resources. It can generate employment opportunities through backward and foreward linkages. The former includes pond preparation, seed production, feed preparation, input (lime, fertilizer, chemicals, drugs etc) marketing, net making, harvesting etc. The latter includes post harvest care, processing, icing, transport, marketing, banking etc. According to Ravichandran (2006) employment generated is 180 man days/ha/crop versus 600 man days /ha/crop in shrimp culture. A labourer can earn Rs7500 per year in agriculture versus Rs12,000/- per year in shrimp culture. So good amount of employment opportunities can be created through shrimp farming. But the main impediment in such development is lack of knowledge and skill of farmers. Extension machinery should well penetrate in such areas and take steps in this direction. In the case of agriculture, only 20% farmers are approached by extension officers and 30% of available technology has reached the farmers. It will be much less in the case of fisheries and aquaculture. According to 11th Five year plan document, only 2% work force in India has skill training versus 75% in Germany, 80% in Japan and 96% in Korea. It indicates that vast majority of our people do not have any identifiable marketable skill as being needed by the society for which persistence of poverty (nearly half of population) is highest in India. There is need to identify suitable potential areas in different localities for shrimp culture and then to identify youth and women self Help Groups (SHG) for the purpose. Then such team can approach nearby Banks with concerned Extension Officers for finance. They must be trained properly to take up farming activities with subject matter specialists. It will benefit such people to earn suitable income and provide substantial employment in remote coastal villages.

UNIO – PEARL CULTURE

Classification

Phylum: Mollusca; Class: Pelecypoda; Order: Eulamellibranchiata; Genus: *Unio*

Introduction

Pearl is a natural germ and is a precious gift of nature. It is a hard object produced within the soft tissue of a living shelled mollusc. The English word pearl comes from the French "Perle", originally from the Latin word "Perna". Pearls are nearly 100% calcium carbonate ($CaCO_3$) and conchiolin. $CaCO_3$ in the form of the mineral aragonite or a mixture of aragonite and calcite held together by an organic horn-like compound called conchiolin. The combination of aragonite and conchiolin is called "nacre", which makes up the "mother of pearl".

Nacre

Nacre is an organic-inorganic composite material produced by some shelled molluscs on the inner shell layer. It is very strong, resilient and iridescent. Inner layer of the shell of many molluscs is also porcellaneous and non-nacreous with usually in a non-iridescent shine. Nacre appears iridescent because the thickness of the aragonite platelets is close to the wavelength of visible light. This results in constructive and destructive interference of different wavelengths of light, resulting in different colours of light being reflected at different viewing angles. Nacre is secreted by the epithelial cells of the mantle tissue of various molluscs. The nacre is continuously deposited onto the inner surface of the shell, the iridescent nacreous layer, commonly known as "mother of pearl". The layers of nacre smooth the shell inner surface and help defend the soft tissue against parasites and damaging debris by entombing them in successive layers of nacre, forming either a blister pearl attached to the interior of the shell, or a free pearl within the mantle tissue. It continues as long as the mollusc lives.

What is Pearl?

Pearl is constituted by $CaCO_3$, organic matrix and water. Truly speaking, it is a by-product of an adaptive immune system-like function. Just like the shell of a mollusc, a pearl is made up of calcium carbonate in minute crystalline form, which is deposited in concentric layers. Therefore, the ideal pearl is perfectly round and smooth. Almost any shelled mollusc can by natural processes, produce some kind of "pearl" when an irritating microscopic object become trapped within the mollusc's mantle fold. But majority of these pearls are not valued as gems. Nacreous pearls are having the commercial importance.

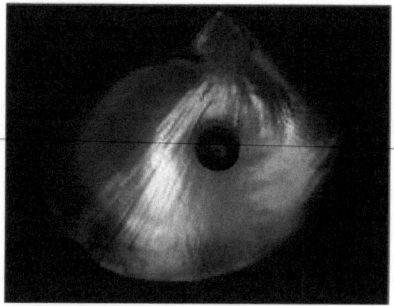
A black pearl and a shell of the black-lipped pearl oyster

The unique luster of pearls depends upon the reflection, refraction and diffraction of light from the transluscent layers. The thinner, and more numerous the layers in the pearl, the finer the luster. The cultured freshwater pearls can be yellow, green, blue, brown, pink, purple or black. But the very best pearls have a metallic mirror-like luster.

Because pearls are made up of primarily $CaCo_3$ they can be dissolved in vinegar. $CaCo_3$ is susceptible to weak acid solution. The acetic acid in vinegar convert the calcium carbonate to calcium, which dissolves in water and the carbonate effervesces as bubbles of carbon dioxide. Nacreous pearls are primarily produced by two groups, molluscan bivalves or clams. Pearl is broadly of 2 types: (i). Natural or Wild Pearl and (ii) Cultured Pearl. One family of nacreous pearl bivalves-the pearl oyster-lives in the sea, while the other- a very different group of bivalves-lives in freshwater, the river molluscs such as freshwater pearl molluscs. Salt water pearls can grow in several species of marine pearl oysters in the family Pteriidae. Fresh water pearls grow within certain species of freshwater molluscs in the order Unionida, the families Unionidae and Margaritiferidae. Natural pearls are near 100% calcium carbonate and conchiolin. Natural pearls form under a set of accidental conditions when a microscopic intruder or parasite enters a bivalve molluscs and settles inside the shell. The molluscs, being irritated the intruder, form a pearl sac of external mantle tissue cells and secretes the calcium carbonate and conchiolin to cover the irritant. This secretion process is repeated many times, thus producing a pearl. Natural pearls come in many shapes, but mostly with perfect round shaped. The build-up of a natural pearl consists of a brown central zone formed by columnar calcium carbonate (usually calcite, sometimes columnar aragonite) and yellowish white outer zone consisting of nacre (tabular aragonite). Cultured pearls are the response of the shell to a tissue implant. A tiny piece of mantle from a donor shell is transplanted into recipient shell. This graft forms a pearl sac and the tissue precipitate calcium carbonate into this pocket. Freshwater or seawater shells can be used as transplant material. The transplant is introduced into the mantle or into the gonad. Besides, a spherical bead can be introduced into mantle. The majority of the saltwater cultured pearls are grown with beads. A beaded cultured pearl shows a solid center with no concentric growth rings, whereas a natural pearl showsa series of concentric growth rings. With X-ray it is possible to see the growth rings of the pearl, where the layers of calcium carbonate are separated by thin layers of conchiolin. A beadless cultured pearl, be it of freshwater or saltwater origin, may show growth rings with a complex central cavity.

Pearl Hunting

For thousands of years, most seawater pearls were retrieved by divers working in the Indian Ocean, in areas like Persian Gulf, Gulf of Mannar. Chinese hunted extensively in South China Sea.

Development of pearl farming

The cultured pearls in the market are of two main categories as beaded cultured pearls and non-beaded freshwater cultured pearls. The first category is gonad grown and usually one pearl is grow at a time. This pearl culture process was first developed by British biologist. The second category is grown in the mantle and up to 25 grafts can be implanted.

The nucleus bead in a beaded cultured pearl is generally a polished sphere made from freshwater mussel shell. Along with a small piece of mantle tissue from another donor molluscs shell to serve as a catalyst for the pearl sac, is surgically implanted into the gonad. In freshwater pearl culture, only the piece of tissue used in most cases and is inserted into the fleshy mantle of the host mussel.

Recent pearl production

The original Japanese cultured pearl, known as akoya pearls are produced by a species of small pearl oyster,*Pinctada fucata martensil*, which are around 8 cm insize. In 2010, China overtook Japan in akoya pearl production. But Japan maintains its status as a pearl processing center, and imports Chinese akoya pearl. In the past two decades, cultured pearls have been produced using larger oysters (*Pinctada maxima*) in the south pacific and Indian Ocean.

Freshwater Pearl farming

In 1914, pearl farmers of Japan began growing cultured freshwater pearls using the pearl mussels native to Lake Biwa, near Kyoto. The extensive and successful Biwa-pearl production was at its peak in1970s, up to six tonns of cultured pearls. But pollution has caused the virtual extinction of the industry. Subsequently, Japanese pearl producers invested in producing cultured pearls in the regions of Shanghai, China. China has since become the world's largest producer of freshwater pearls, producing more than 1500 metric tons per year.

Culture Practices

Farming practice of the freshwater pearl culture operation involves six major steps sequentially viz., collection of mussels, pre-operative conditioning, surgery, post-operative care, pond culture and harvesting of pearls.

Collection of mussels

The healthy mussels are collected from the freshwater bodies like pond, river etc. They are collected manually and kept in buckets or containers with water. The ideal mussel size used for pearl culture is over 8 cm in anterior-posterior length.

Pre-operative conditioning

The collected mussels are kept for pre-operative conditioning for 2 to 3 days by keeping them in crowded condition in captivity with aged tap water at a stocking density of 1 mussel/liter Pre-operative conditioning helps in weakening of adductor muscles, which helps in easy handling during surgery.

Mussel surgery

Depending on the place of surgery the implantation is of three types, *viz.* mantle cavity, mantle tissue and gonadal implantations. The key materials required during the surgical implantations are beads or nuclei, which are usually made from mollusc shell or other calcareous materials.

Mantle cavity implantation: In this procedure round(4-6 mm diameter) or designed (images of Ganesh, Buddha, etc.) beads are inserted into the mantle cavity region of mussel after opening the two valves (without causing injury to mussels at both ends) of animal and separating carefully the mantles of anterior sides from the shell with the help of surgical set. Implantation could be done in mantle cavities of both the valves. In case of implantation of designed beads care is taken such a way that the design portion faces the mantle. After placing the beads in desired place the gaps created during implantation are closed just by pushing the mantle onto the shell.

Mantle tissue implantation: Here the mussels are divided into two groups; the donor and the recipient mussels. The first step in this procedure is preparation of graft (small pieces of mantle tissue). This is done by preparing a mantle ribbon (a strip of mantle along the ventral side of the mussel) from a donor mussel, which is sacrificed, and cutting that into small pieces (2 x 2 mm). The implantation is done on recepient mussels, which is of two types *viz.*, non-nucleated and nucleated. In the former, only the graft pieces are introduced into the pockets created at the inner side of posterior pallial mantle present at the ventral region of the mussel. In the nucleated method, a graft piece followed by a small nucleus (2 mm dia) is introduced in the pockets. In both the procedures care is taken so that graft or nucleus does not come out of the pocket. Implantations could be done at mantle ribbons of both valves.

Gonadal implantation: This procedure also involves preparation of grafts as described earlier (mantle tissue method). First a cut is made at the edge of the gonad of the mussel. Then a graft is inserted into the gonad followed by nucleus (2-4 mm dia) so that the nucleus and graft should be in close contact. Care is taken such a way that nucleus touches the outer epithelial layer of the graft and the intestine is not cut during the surgery.

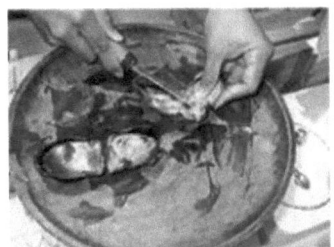

A pearl being extracted from an akoya pearl oyster

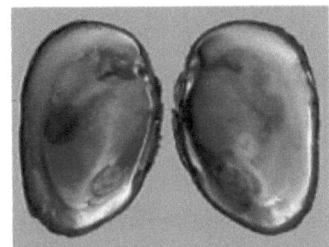

Shell of one species of freshwater pearl mussel, *Margaritifera margaritifera*

Post-operative care

Implanted mussels are kept in post-operative care unit in nylon bags for 10 days with antibiotic treatment and supply of natural food. The units are examined daily with removal of dead mussels and the ones that reject the nucleus.

Pond culture

Freshwater pearl mussel culture

After post-operative care the implanted mussels are stocked in the ponds. The mussels are kept in nylon bags (2 mussels per bag) and are hung from bamboo or PVC pipes and placed in ponds at 1 m depth. The mussels are cultured at stocking density of 20,000-30,000/ha. The ponds are fertilized with organic and inorganic fertilizer periodically to sustain the plankton productivity. Periodical checking of mussels with removal of dead ones and cleaning of bags is required throughout the culture period of 12-18 months.

Pearl harvest

Collection of round cultured pearls

At the end of the culture period the mussels are harvested. While individual pearls can be taken out from the mantle tissue or gonad of the live mussels, the mussels are sacrificed in case of mantle cavity method. The products obtained through different surgical implantation methods are shell attached half round and shell attached image pearls in mantle cavity method; unattached small irregular or round pearls in mantle tissue method; and unattached big irregular or round pearls in gonadal method.

Points to be considered:

The following projections are based only on the experimental results obtained at CIFA, Bhubaneswar.

The design or image pearl is an old concept, however, the array of design pearls produced at CIFA are having considerable fancy value in the context of huge imports of Chinese semi-cultured, 'rice pearls' in the domestic market. The variable charges like consultancy and marketing are not included in the calculations.

Operational details

Area	:	0.4 ha
Product	:	Design pearls through double implantations
Stocking density	:	25,000 mussels/0.4 ha
Culture period	:	One and half year

Sl. No.	Item	Amount (Rupees in lakhs)
I.	**Expenditure**	
A.	**Fixed Capital**	
1.	Operation shed (12 m x 5 m)	1.00
2.	Mussel holding tanks (20 ferro-cement/FRP tanks of 200 l capacity @ Rs.1,500/tank)	0.30
3.	Culture units (PVC pipe and floats)	1.50
4.	Surgical sets (4 sets @ Rs 5,000/set)	0.20
5.	Furniture for surgical facilities (4 sets)	0.10
	Sub-total	**3.10**

B.	Variable Costs	
1.	Pond lease value (for 1 1/2 year crop)	0.15
2.	Mussels (25,000 nos @ Re 0.5/ mussel)	0.125
3.	Design pearl nuclei (50,000 nos for double implantation @ Rs 4/nucleus)	2.00
4.	Skilled workers for implantation (3 persons for 3 months @ Rs. 6,000/person/month)	0.54
5.	Wages (2 persons for 1½ years @ Rs 3,000/person/month for farm maintenance and watch & ward)	1.08
6.	Fertilisers, lime and other miscellaneous costs	0.30
7.	Post-harvest processing of pearls (9,000 design pearls @ Rs.5/pearl)	0.45
	Sub-total	4.645
C.	Total Costs	
1.	Total variable costs	4.645
2.	Interest on variable cost (@15% half yearly)	0.348
3.	Depreciation cost on fixed capital (for 1 ½ years @ 10% yearly)	0.465
4.	Interest on fixed capital (for 1 ½ year @15% per annum)	0.465
	Grand Total	5.923
II.	Gross Income	
1.	Returns on sale of pearls (30,000 pearls from 15,000 harvested mussels considering 60% survival)	
	Design pearls (Grade-A 10% of total) 3,000@ Rs 150/pearl	4.50
	Design pearls (Grade-B) (20% of total) 6,000 @ Rs. 60/pearl	3.60
	Gross Return	8.10
III.	Net Income (Gross Income - Total Costs)	2.177

White pearl necklace

Freshwater pearl culture can fetch a good return and can provide livelihood to the farmers. It can also solve unemployment problem to some extent. As the demand of pearl is continuously increasing, along with its increase in the import load, this has been affecting the economy of the country. Government should encourage the farmers by providing them suitable training for better harvest.

FISH – PISCICULTURE

Introduction

In Country like India the intake of meat and milk is low, so fish has special importance as a supplement to ill-balanced cereal diets. Today protein deficiency is the world's most serious human malnutritional problem, and perhaps 30 to 40% of the world population is suffering from protein deficiency. It is estimated that about 8.5 million tons of fish is required annually to meet the present day demand of fish protein in the country against an annual production of only 1.7 million tons.

In India, inland water with potentialities of fish culture is approximately 7.5 million hectares or 2.34% of the total area of the country. Many of the water reservoirs remain either unused or not properly used for fish culture for want of proper scientific know-how. In recent years researches conducted by the Central Inland Fisheries Research Institute have revolutionized fish culture in India and a net production of 85,000 kg/ hectare/year has already been achieved. India has 8,118 kilometers of marine coastline, 3,827 fishing villages, and 1,914 traditional fish landing centers. India's fresh water resources consist of 195,210 kilometers of rivers and canals, 2.9 million hectares of minor and major reservoirs, 2.4 million hectares of ponds and lakes, and about 0.8 million hectares of flood plain wetlands and water bodies. As of 2010, the marine and freshwater resources offered a combined sustainable catch fishing potential of over 4 million metric tonns of fish. In addition, India's water and natural resources offer a tenfold growth potential in aquaculture (farm fishing) from 2010 harvest levels of 3.9 million metric tonns of fish, if India were to adopt fishing knowledge, regulatory reforms, and sustainability policies adopted by China over the last two decades.

As of 2010, fish harvest distribution was difficult within India because of poor rural road infrastructure, lack of cold storage and absence of organized retail in most parts of the country.

Fish culture received notable attention in Tamil Nadu (formerly the state of Madras) as early as 1911, subsequently, states such as West Bengal, Punjab, Uttar Pradesh, Gujarat, Karnataka and Andhra Pradesh initiated fish culture through the establishment of Fisheries Departments. In 2006, Indian central Government initiated a dedicated organization focussed on fisheries, under its Ministry of Agriculture.

India laid the foundation for scientific carp farming in the country between 1970 and 1980, by demonstrating high production levels of 8 to 10 tonns/hectare/year in an incubation center. The late 1980s saw the dawn of aquaculture in India and transformed fish culture into a more modern enterprise. With economic liberalization of early 1990s, fishing industry got a major investment boost. India's breeding and culture technologies include primarily different species of carp; other species such as catfish, murrels and prawns are recent additions. The different culture systems in Indian practice include.

- Intensive pond culture with supplementary feeding and aeration (10–15 tonns/ha/yr)
- Composite carp culture (4–6 tonns/ha/yr)
- Weed-based carp polyculture (3–4 tonns/ha/yr)
- Integrated fish farming with poultry, pigs, ducks, horticulture, etc. (3–5 tonns/ha/yr)
- Pen culture (3–5 tonns/ha/yr)
- Cage culture (10–15 kg/m²/yr)
- Running-water fish culture (20–50 kg/m²/yr)

Carp hatcheries in both the public and private sectors have contributed towards the increase in seed production from 6321 million fry in 1985–1986 to over 18500 million fry in 2007.

Table 1. Leading fish producing states in India, 2007– 2008

Rank	States	Total production (metric tones
1	West Bengal	1,447,260
2	Andhra Pradesh	1,010,830
3	Gujarat	721,910
4	Kerala	667,330
5	Tamil Nadu	559,360
6	Maharashtra	556,450
7	Odisha	349,480
8	Uttar Pradesh	325,950
9	Bihar	319,100
10	Karnataka	297,690

Aims of Fish Culture
- The main aim of fish culture is to obtain maximum yield of fish.
- To obtain palatable and highly nutritive fish flesh.
- By-products of fishing industry.

Qualities of Culturable Fishes

The main attraction of fish culture is to obtain more fish. So, before starting this complicated programme it is essential to have a comprehensive idea about the fish itself. The culturable fishes.

1. should have ability to feed on natural food.
2. should have ability to feed on artificial diet.
3. should consume small quantity of food for development.
4. should be herbivorous in nature.
5. should be able to tolerate a sudden change in climatic conditions and nature of water in ponds.
6. should be fast growing and can attain good length and weight both.
7. should live in ponds with other fishes without any disturbance
8. should be able to resist against diseases.
9. should be prolific breeder.
10. should be palatable and much nutritive.

Types of Cultivable Fishes

The cultivable fishes are of 3 types:

1. Indigenous or native fresh water fishes *viz.,* major carps.
2. Salt water fishes acclimatized for fresh water viz., Chanos, Mullets.
3. Exotic fishes, imported from other countries *viz.,* Mirror carp, Chinese carp, Crucian carp and Common carps.

Types of Culture in India

It is practically difficult to find all the characteristics of a culturable fish in any one of the fishes but some fish are found most suitable for fish culture. Among these 'Major Carps' have proved to be best culturable fish in India having following qualities.

- Carps feed on zoo-and phyto-planktons, decaying weeds, debris and other aquatic plants.
- Carps can survive under somewhat higher temperatures and also in turbid water.

- Carps can tolerate oxygen variation in water.
- Carps have fast growth rate and can attain good size and weight.
- Carps can be transported from one place to other easily.
- The flesh of carps is mostly palatable and much nutritive.

External Factors Affecting Fish Culture

Now-a-days, profitable fish culture has become purely technical and scientific because a munber of environmental factors affect the fish culture programme. A number of factors *viz,,* temperature, light, rain, water, flood, water current, diseases, toxic pollutants, pH, hardness and salinity of water, dissolved oxygen etc, play a vital role in fish culture. The temperature and light intensity both have profound effect because fish cannot breed above or below the critical temperature which may be variable according to the species of fish.

Management of Fish Culture Programmes

Fish culture is a complicated process, so for an ideal fish culture one should have an idea about the different stages of fish culture *i.e.* topographic situation, quality of water, source of water, and other physical, chemical arid biological factors. Ponds are the sites where fish develop and grow. The management of ponds has its own importance and has to be tackled from the point of view of breeding, hatching, nursing, roaring and stocking ponds. The nature of ponds may slightly vary with the species on the fish. Even different stages of the same fish are cultured in the ponds having quite different properties. Keeping in view the various stages of fishes, the following different types of ponds have been recommended to manage them.

Breeding Pond

First step in the fish culture is the breeding of fishes, therefore, for proper breeding special types of ponds are prepared called as breeding ponds. These ponds are prepared near the rivers or other natural water resources.

Types of breeding

According to the mode of breeding there are two categories:
   (1) Natural breeding (Bundh breeding).
   (2) Induced breeding.

Natural breeding (Bundh breeding)

The natural bundhs are special types of ponds where natural riverine conditions or any natural water resource conditions are managed for the breeding of cultivable fishes. These specially designed bundhs are constructed in large low-lying area having facility to accommodate large quantity of rain water. These bundhs are having an outlet for the exit of excess rain water. The shallow area of such bundhs is always used as spawing ground. These bundhs are of three types:

(1) Wet bundh   (2) Dry bundh   (3) Modern bundh

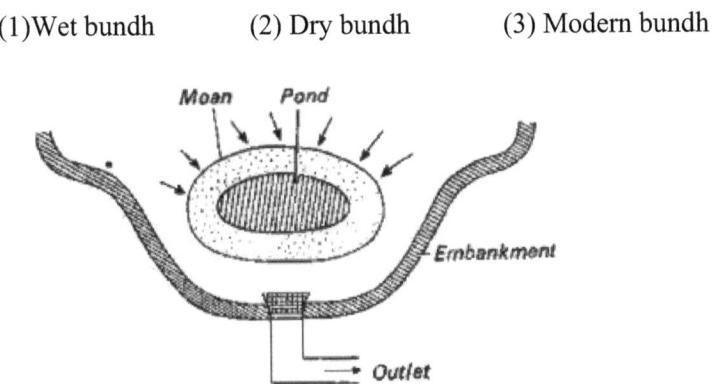

1. **Wet bundh.** The ponds specially constructed for fish breeding having water throughout the year are known as wet bundhs or perennial bundhs.

    An inlet is formed at the higher level of bundh for the entrance of the water while an outlet is prepared in low lying area for the exit of the water from the bundh. The flow of water from outlet is controlled with the help of bamboo fencing.

2. **Dry bundh.** This type of pond is purely seasonal with shallow water areas. This is constructed by keeping soil walls from three sides and open area from one side. In monsoon period rain water flows towards this bundh and fills the pond. But after monsoon water this bundh dries up after a month or two.

3. **Modern bundh.** This is known as 'Pucca bundh'.

    It is a masony construction and a sluice gate at the lower-most level of the bundh is the characteristic feature. The total exit of water from the bundh is possible by this gate so that after each spawning, bundh is cleared of water.

According to the breeding nature of different fishes, suitable bundhs are used for spawning.

Induced breeding

The fish seed is commonly collected from breeding grounds but this has certain drawbacks because a mixture of eggs of various types of fishes is obtained in which some eggs may be of predatory fishes. Keeping this in mind some advanced techniques have been devised for the super quality of fish seed by artificial method of fertilization.

Artificial fertilization. In this method of fertilization ova from the females and the sperms from the males are taken out by artificial mechanical process and the eggs are got fertilized by the sperms. This is of two types:

(1) Dry method   (2) Wet method

1. **Dry method.** Eggs and sperms for fishes (which are just taken out) are mixed thoroughly and left in this condition for about 30 minutes. Now water is added in this mixture after the fertilization of eggs.
2. **Wet method.** Eggs are kept in water and milt of the male is spread directly into this water. This method is commonly used in sticky eggs.

For artificial fertilization, mature female fish having eggs is taken in hand and dried by towel. The belly of this fish should be upward. Stripping should be started with the thumb of the right hand from the anterior to the posterior direction for the ejection of the eggs due to force. In this way eggs are collected separately. Further, a male fish is caught and its belly-is kept down-wards. The milt of the fish is stripped with the help of the thumb and index finger and collected separately. Now the eggs are subjected to fertilization as mentioned above.

Types of induced breeding. In accordance with the breeding habit, two types of fishes are found i.e., fishes commonly breeding in ponds and fishes not commonly breeding in ponds. Therefore, induced breeding is managed in both types of fishes by different methods as described below.

1. Method of inducing in pond breeding fishes.

Although pond breeding fishes are well adapted for breeding in the maintained ponds yet for effective and successful breeding some inducement is given to thepond breeding fishes also. First of all they are segregated sex-wise to prevent wild spawning. Further, they are allowed to breed under the inflitence of any one of the following spawning stimulants:

(i) by the introduction of specific selected brood fishes in breeding ponds. (ii) by replacement of old water by fresh rain water. (iii) by providing suitable site for the attachment of the laid eggs. (iv) by providing particular temperature and light intensity.

2. Method of inducing in difficult spawning fishes.

The fishes like Chinese carp, Cat fishes, Mullets, Milk fish etc. do not commonly breed in ponds, therefore, some breeding stimuli are needed for proper breeding which are given below:

(a) Induced bundh breeding. Spawning in bundhs occurs after continuous heavy showers when rain water covers the bundh. The breeding of carp is very much interesting. The male and the female indulge in sex play. The male starts chasing the female with vigorous splashing of water. The-female is then held by the male and ova are released, at the same time male ejects the milt over the eggs. After spawning the breeders remain guilt for some time and then swim away. The bundh breeding of 'hardly spawning' is induced by the following factors:

1. Spawning is stimulated by heavy monsoon.
2. Water current also induces spawning.
3. Temperature between 24 to 32°C is found best inducer for spawning of fishes.
4. Cloudy season followed by thunder, storm and rain also stimulates spawning of fishes.

(b) Induced breeding by hypophysation (Hormone)

The gonadotropin hormone (F.S.H. and LH) secreted by pituitary gland influences the maturation of gonads and spawning in the fishes. In India, Khan (1938) successfully induced *Cirrhinus rnrigala* to spawn by injecting mammalian pituitary hormone.

Method of hypophysation: First of all pituitary is takenout, then preserved in absolute alcohol inside the sealed tube in a dessicator at room temperature or in acetone for about 36 hours and stored in sealed phails in refrigerator. The pituitary glands should'be taken out from fully mature, healthy and freshly killed fishes. The donor fish of the same species are most preferred for this purpose.

For the preparation of extract, the weighed glands are homogenized in distilled water or 0.3% saline or glycerine and centifuged for 15 minutes at 20,000 rpm. The supernatant thus obtained is injected intramuscularly on the back or on the base of the caudal fm. In some cases intraperritoneal injection is also recommended at the base of the pectoral fin.

Dose for hypophysation: For the first time 2 to 3 mg of pituitary extracts/ kg body weight of fish is injected. After 6 hours of first injection, slightly higher dose of 5 to 8 mg/kg body weight of female and 2 to 3 mg/kg body weight of male are injected. Now both male and female are released into a 'cloth hapa' which is fixed with bamboo poles in the breeding pond. The breeding hapa is a rectangular box formed of mosquito net cloth and closed from all sides. After 24 hours of injection the male and the female start to swim, become excited and restless and always try to jump out of the hapa. The male and female chase each-other and push each-other with their snouts and spawning occurs as result of which eggs are fertilized. Now the fertilized eggs are removed from spawning place and kept into hatching hapas.

Synthetic hormone: With the advancement of fishculture programme the synthetic hormones have been introduced in the field for breeding of fishes. The human chorionic gonadotropin (HCG) and some cortic steroids particularly 'Deoxycorticosteron' are most successfully used for induced spawning of culti-vable fishes. A mixture of chorionic gonadotrophic and mammalian pituitary extract as 'Synaphorin' is used in combination with the fish pituitary extract.

Fish Seed

The fish seed is collected from the breeding ponds, a large number of fish collection stations have been established at Ganga, Yamuna and Brahamputra etc. The fish seeds are collected from the breeding grounds of Ganga, Yamuna, Gomati, Betwa, Ghaghra and others: The best places for collection of fishseeds are on the curve area of these rivers. A tailed-structure is attached to the whole net which is a rectangular piece (2' x 1' xr) called as Gamchaa. All stages of fish can be collected in this gamchaa from where they are scooped and transferred to the 'Hatching pits'.

Hatching Pit

The fertilized eggs are kept into hatching pits for hatching. At the time of construction of hatching pits one should see that the hatching pit :

(1) should be nearer to the breeding grounds.

(2) should be smaller in size.

(3) should contain such a quantity of water which must dry within a month or two.

(4) should be more in number.

Types of hatching pits

Hatching pits are of two types:

1. Hatcharies. These are small sized ponds (Fig.2) in which fertilized eggs are transferred. After 2 to 15 hours the fertilized eggs are hatched. Some draw backs make the hatcharies unfit for advanced fish culture programme. These drawbacks are as under:

(i) sudden rise and fall in temperature, (ii) entrance of predators in ponds, (iii) drying of water from ponds may cause mortality of eggs.

To overcome all these drawbacks specially designed hatcharies and hatching hapas are constructed.

2. Hatching hapas. Hapas (Fig.3) arerectangular trough shaped tanks made up of cloth supported by bamboo poles fixed in the river. In these hapas fish eggs are aerated by continuous flow of current. The size of hapa is about 3' x1'5' x and is made up of mosquito net cloth which is fixed into outer larger hapa made up of coarse cloth. Two types of hapas are designed:

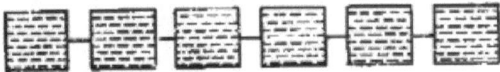

Fig. 2. Hatcharies

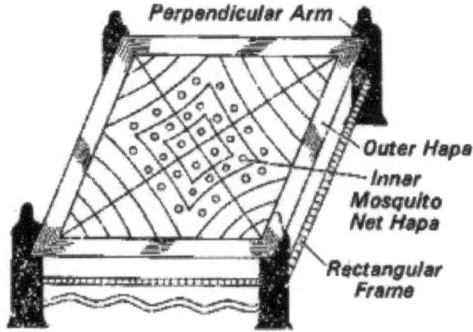

Fig.3. Hatcharies hapas

(a) *Fixed type hapa* .If possible to fix the perpendicular poles (arms of hapa) only then the fixed type of hapa is used for hatchings of eggs otherwise floating type of hapas should be used.

(b) *Floating type hapa.* In hard bottom areas whereit is not possible to use fixed hapas, floating hapas are built. These hapas are arranged in large number in series attached with the bamboo and floated on water surface. To avoid over crowding of eggs, only one layer of eggs is spread in each hapa. Hatching of eggs takes place in the outer hapas leaving the egg membrane in the inner one. The hatchlings are kept for 36 to 48 hours in hapas and then transferred to the nurseries.

Transport of Fish Fry to Nursery Ponds

The fish fries are collected and transported to nurseries. In West Bengal these are transported in earthen 'Hundies' which cause heavy mortality due to the following factors.

(1) Decreased dissolved oxygen concentration in water.

(2) Increased carbondioxide concentration in water.

(3) Toxicity of wastes like excreted ammonia.

(4) Hyperactivity and its strain.

(5) Physical injury to fries during transport.

To overcome the above mentioned injuries, fries are subjected to conditioning before transportation by keeping them into fixed volume of water for definite period and then transported in open or closed vessels. Temperature of the water of the vessels can be kept lower by tieing a wet cloth around the vessel. Dead fries should be removed to avoid the pollution and infections. Now-a-days fries are transported in sealed metal containers with oxygen. In India alkethen bags of different sizes are used for the transportation of fish fries from hatching hapas to nursery-ponds.

Nursery Pond

The newly,hatched fries transported from hatching hapa to nursery ponds are very tender so one should be very careful in maintaining them there. Nursery ponds should always be near the hatching hapas. The nursery ponds are small set of shallow (3' to 5' in depth) water reservoir. But now-a-days generally more deep (5' to 8') ponds are prepared having an area of about 1/2 acre. The ideal and recommended nursery ponds are 50 to 60' x 30 to 40' x 4 to 5'.

The nursery ponds should be prepared before the hatching of fries. The exit and entrance of water should also be under control.

In nursery ponds natural resources of food are less and when large number of fries are released they certainly suffer due to lack of food. First of all predatory and weed fishes should be removed from pond. The chemical fertilizers viz., ammonium sulphate, sodium nitrate and superphosphate should be used along with cowdung . The amount of chemical fertilizers should be less in quantity. 5,000 to 70,000 kg cow dung/hectare is recommended in accordance with the nature and fertility of the pond soil. Due to these manures numerous zooplanktons are developed within 10 to 20 days and phyto planktons are also grown up within 10 to 15 days. On these phyto- and zooplanktons they can feed easily and comfortably.

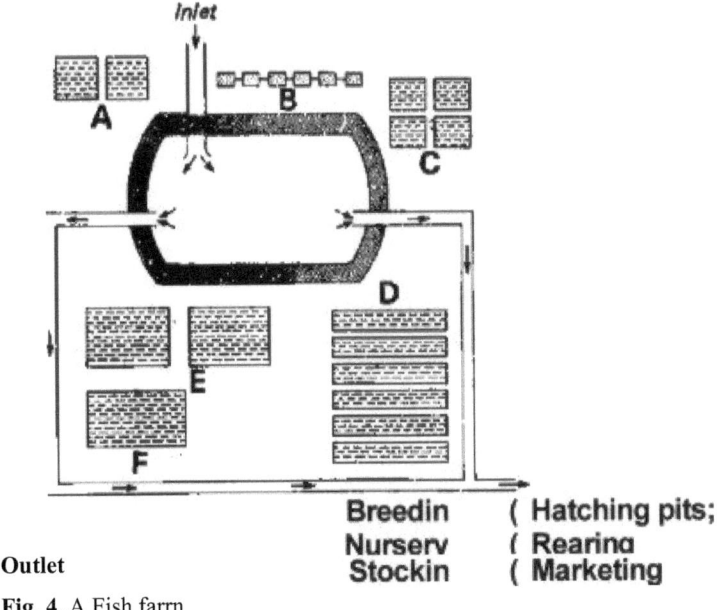

Outlet

Breedin (Hatching pits;
Nursery (Rearing
Stockin (Marketing

Fig. 4. A Fish farm

Mortality in nursery ponds

The heavy mortality of fish fries has been recorded in nursery ponds. The following factors are responsible for such mortality:

(1) Sudden change in the quality of water form hatching hapa to nursery ponds.

(2) Lack of suitable food in pond.

(3) Presence of predatory fishes and predatory aquatic insects in, the pond.

(4) Overgrowth of plankton.

(5) Decreased oxygen concentration in water,

(6) Cannibalism.

Precautions for nursery ponds

(1) In the nursery ponds water should be under good control and circulating.

(2) Pond should be nearer to the hatching ponds.

(3) Ponds should be predator free.

(4) To avoid the overcrowding, fries should be kept in limited number.

(5) Supply of food material should be proper.

When fries are more developed in the nursery ponds and attain a length of 10 to 15 cm they should be transferred into the rearing ponds.

Rearing Ponds

For good health and growth of fingerlings the exercise is essential for them inside the rearing ponds. So these fingerlings are reared in longer and narrower ponds to provide them long distance for swimming. The water of this pond may be seasonal or perennial. The rearing ponds should be free from toxicant and predators. The depth of the pond should be about six feet containing nutritive food material in accordance with the population of fingerlings. As fingerlings attain a length of about 20 cm they should be transferred to the other type of ponds called as stocking ponds.

Transport of Fingerlings

The fingerlings are transported form rearing ponds to stocking ponds in a container of 1,000 litre capacity. This container is internally lined with foam to 'avoid physical injury. The proper arrangement for aeration of the tank is essential during transport. To make the fingerlings inactive various sedative (sodium amylate and barbifurate) are used. This is only for less consumption of dissolved oxygen during transport. Sometimes some diseases, parasites and predators are also transported alongwith fingerlings.

To avoid these, fingerlings should be washed carefully before packing and the use of antibiotics, methyl blue, copper sulphate, potassium permanganate, formalin, common salt are recommended. Taking all these precautions, the finger-lings are transferred to the stocking ponds.

Stocking Ponds

The stocking ponds should be cleaned of weeds and predatory fishes. Sufficient food is essential for good growth of fishes in these ponds hence proper manuring should be done to increase the productionof zooplankton because large number of fishes may not be able to feed properly due to lack of food material. As for the proper organic manuring cow dung is the best and should be used at the rate of 20,000 to 25,000 kg/ hectare/year. The inorganic, chemical fertilizers are also used *viz.,* super-phosphate, ammonium nitrate and ammonium sulphate at the rate of 1,000 to 1,500 kg/ hectare/year.

The powdered rice, paddy, oil cakes, coconut, mustard, groundnut, etc., are commonly used as artificial food for the fishes. The artificial food used for the fishes should be easily digestible in natural form and economically suitable. The best time for feeding the fishes is in the morning hours. The quality of food should not be changed suddenly. The amount of fertilizers used is totally dependent on the fertility of the soil, number of fishes and types of fishes being kept in the stocking ponds. When fishes attain maximum length and weight they should be harvested.

Harvesting

Harvesting is done to capture the fishes from the water. The well grown fishes are taken out for marketing and smaller ones are again released into stocking ponds for their growth. In highly organised and well planned fish farming the fishes below a particular size are not generally captured.

Methods of Fishing

For proper and scientific management of fishing from different resources, the idea on fishing gear and net in accordance with the size and habitat of fish is essential. The oldest art of fishing in the form of stones and spears has been converted into complex traps, nets and lines. In our country, though a large number of fishing methods are in use but all are not of common use throughout the country. Some common methods of fishing are described here:

1. **Stranding**. It is widely used in shallow water resources in which some of the shallow area of the pond is separated from the main body by throwing up a low earthen bundh. When almost all the water is thrown out from this area, fishes are gathered and caught out. This method is economical but can only be used in shallow swains and burrow pits.

2. **Angling**. This is also very common in practice for fishing of large and predatory fishes. Along the banks of rivers or ponds fishing rods are left dug into the soil and the owner of rod visits to unhook the occasional catch or renew the bait consisting of earthworms. Sometimes very long lines containing 100 to 250 hooks attached to it, are also used for fishing. This method can be used both in shallow and deep waters.

3. **Traps.** This may also be used in shallow and deep waters both. Traps are made of basket work of varying materials. They vary in shape, size and mode of operation. The cage trap is of fixed type whereas scoop and cover traps are movable at the time of operation. Cover traps are under common practice throughout the country and specially used in muddy bottom. The basket is plugged into the water where fish is supposed to be present and then the fish is searched by hand from an opening on the top of the trap. Fishes like *Clarias, Heteropneustes, Anabas, Eel* etc. are trapped commonly. Sometimes in deeper water two persons are needed for the operation of the traps.

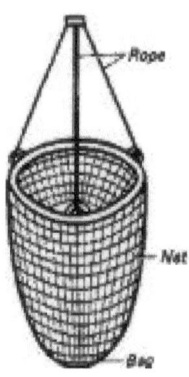 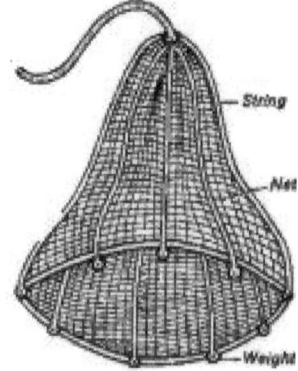

Fig.5   Fig.6

4. **Scooping.** The rim of rectangular, triangular or round bamboo frame is used and the net of same size and shape is loosely hung with this frame. The rounded net is generally used for transfer of fresh catch from the drag net.

5. **Dip of lift net**. The basic principle of lift net fishing is to keep the net in water for about 5 to 10 minutes and then pull it out rapidly. In this method the fishes which are over the net at that time may be caught. Some baits are also put on the net or suspended over it. On this principle varying nets of different shapes and sizes are used in different regions. All these nets are fixed to frame for catching large and small sized fishes.

6. **Cast net (Ghagaria jal)**. This is a circularumbrella-shaped net made up of cotton twine. The number of meshes on the mouth is 50 and at the periphery around 1,000. The size of the mesh is 2-5 cm and at the periphery circular pockets are formed by folding the net inwards to about 61 meshes in depth. Each pocket contains two iron sinkers. On its apex a thin rope of 3 mm diameter is woven through 50 meshes and it is tightened to close the mouth. The free end of the rope is tied to the fore finger of the left net. At the time of fishing the net is swung over head and dropped to the distance reachable by attached rope Cast net is operated from the boat. After sometime it is slowly dragged in water with the rope. The fishes enter in the pockets and thus are caught and collected (Fig. 6)

Purse net. Purse nets are generally used tocapture the migratory fishes like *Hilsa* in the month of October to January and large sized carps and cat fishes are fished in May to July. In river Ganges purse net fishing is recorded. Two types of purse nets are operated in our country i.e., Kharki Jal and Shangla Jal. The wide mouth has two flexible bamboo rods hinged at the two angles and forming upper and lower lips. The net is suspended in deep water and attached from the boat, At the time of operation the mouth is kept open but as fish enter into the net, mouth is closed due to pressure on the bamboo poles (Fig. 7).

Gill net. Gill nets are wall-like with a mesh opening of varying size with type of fish to be fished. The net is set. on the transverse direction of the migrating fish. The net is made up of common hemp fibers but now-a-days nets are prepared from synthetic impermeable fibers due to which it is not possible for the fishes to see the nets. So, as fishes try to swim through a net wall the meshes form a noose round its head and they are caught. Three types of gill nets are commonly used viz., Set net (stationary nets), Floating gill net and drift net. The net is set in the morning hours and caught fishes are collected in evening. The fishes caught by this net are *Seenghala, Pangasius*, Major carps, *Silonia,* etc. This net is fit for fishing in fast current (Fig. 8).

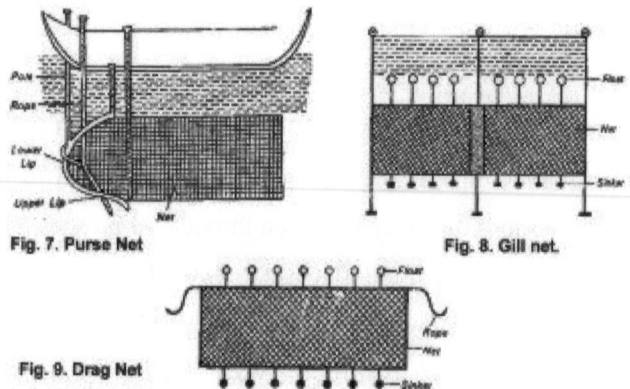

Fig. 7. Purse Net Fig. 8. Gill net.

Fig. 9. Drag Net

7. **Drag net.** This net is commonly used for fishing in lakes and rivers. It is provided with a main rope which carries floats and a foot rope bearing sinkers. The shape and size of the net is varying according to the water source where fishing is to be performed. Some drag nets are provided with peripheral pockets (Fig. 9).

Mainly two types of drag nets are constructed:
- The first one is consisting of a bag with two wings.
- The second is a wall like formed of a very long net of which the upper margin is supported by a strong rope and having a number of wooden floats. It extends from bank to bank and needs 25 to 30 fishermen to drag it.

It is hoped that expert and advanced technology would be able to design some net in future which may make fishing efficient and easy.

Electric Fishing

Although it is not of common use but sometimes it is used with precautions for fishing. It is reported that fishes are attracted towards electric field on electrode. The current is provided by generator or from batteries. The electric current is of about 15 amperes. The voltage may vary according to the depth, distance and conductivity of water. Large fishes are more sensitive than smaller ones. The voltage is between 150 to 250 volts. Direct current, alternating or interrupted current can be used. If the voltage and intensity is well regulated the fish soon recover from the shock in the electric field.

Preservation of Fish

If not preserved, fishes get spoiled by decomposition. Fish flesh comprises of protein, fat, minerals, vitamins, amino acids, iodine, phosphorus and large quantity of water. Fishes are also provided with a number of bacteria which after the death, start to attack various constituents. In this light the preservation of fishes is an essential process. The following methods of preservation of fishes are generally practised in India and abroad.

1. **Refrigeration.** The basic idea behind refrigerationis to preserve the fish at 0°C which prevents the spoilage for short period. For this purpose alternate layers of fish and ice are kept in closed vessels to maintain the temperature at 0°C. In case of large fishes ice pieces are kept in abdominal cavity of gutted fishes.

2. **Deep-freezing.** Before this freezing, fishes arewashed properly and kept at a temperature of —18°C for longer period. Only the fresh fish in good condition are deep-frozen. Before keeping the fish in this process the heads of large fishes are removed. They are also gutted and washed. By this method fishes can be kept for longer Period without spoilage.

3. **Freeze drying.** This is long process and expensive so only good quality of fishes are put to this type of preservation. The first step is the freezing of the fish, which is then allowed to dry by sublimation. In this process ice is changed into water vapour without melting. The colour and nutritive substances are completely preserved by this technique. Further, the fish is frozen to —20°C by keeping them in freezing chamber. After freezing, fishes kept in trays are sent to the cabin containing horizontal heating plates for drying in vacuum. Now fishes are well dried due to hot plates and packed in air-Conditioned chamber.

4. **Sun drying.** Small sized fishes are dried in sunlight especially in tropical countries like India, China, Japan and others, where sun rays are very powerful (scorching) to dry the fishes easily. The fishes are kept for dehydration on a mat or anything for 3 to 5 days and during this period turning over the fishes is continued. Larger fishes are cut into pieces for easy drying. But this method is not perfect for longer preservation.

5. **Sun curing.** An advanced method over simple sun drying is developed in which body of the fish is opened from the ventral side and the viscera and the gills are removed. Then the fish is washed and salted in ratio of 1: 3 to 1: 8 (salt: fish) which is related with the size of fish.

6. **Mona curing.** This is basically similar to suncuring with the difference that no incision is made in the body of the fish to remove the intestine and the gills. In this method the intestine and the gills are removed directly from the mouth. Further, eviscerated fishes are cleaned, salted and dried as earlier.

7. **Wet curing.** It is just like sun curing with the only difference in the packing fish as such. This method is employed only for fatty fishes.

8. **Salting.** The partial dehydration of the fishes by osmosis with sodium chloride is termed as salting. If intense, this will kill the microbes and stop diastasis. The heads of the fishes are removed, gutted and washed then salted as soon as possible. Salting can be carried out as follows :
    (1) Dry (alternate layer of fish and salt).
    (2) In brine (light, 16% salt : strong, 25% salt),
    (3) In brine after the fishes have been dipped in salt.
    (4) Cold by spreading salt and crushed ice on thefish.

For light and cold salting, the processes should be performed in cold room (2 to 3°C) but for strong salting this process can be performed at the normal room temperature.

9. **Smoking.** The preservation by the action of wood smoke is known as smoking of the fish. This permits the preparation of delicate specialities. The temperature of the smoke and its rate of circulation should be controlled. The smoking may be of hot or cold type. For cold smoking fishes are dried, salted, exposed to a smokeless fire (38°C) and then processed for real smoking at 28°C. The hot smoking is performed on fresh fishes in which fish is subjected at 130°C on. strong fire which is followed by smoking at 40°C. The industrial smoking done in galleries with a smoking installation and a system for the proper circulation of smokes. This method was used to preserve the fishes in World War II but it is not accepted in the present day fish industries.

10. **Canning.** The canning of fish is a long,complicated and advanced process in which head and viscera are first removed, eviscerated fishes are treated with brine, washed, dried and cooked in olive oil to remove the excess of water for 2 to 4 minutes only. Now cooked fishes are packed in olive oil in tins and sealed and then despatched out

for markets. This is very costly process so it is not in common use. This process of preservation is widely used in America, France, Japan and Spain.

Composite Fish Farming

It is found that if few selected species of fishes are stocked together in proper proportion in a pond, total production of fishes is increased many times. This mixed fish farming is termed as composite fish farming or Polyculture. The main idea behind this type of farming is :

(1) All available nitches are fully utilised.
(2) Compatible species do not harm each other.
(3) No competition among different species is found.
(4) Fishes may have beneficial effect on each other.

Food material supplied in this type of farming is in accordance with the feeding habit of the different species. In India mixed fish culture is an old practice in which *Catla catla, Labeo rohita* and *Cirrhina mrigala* are surface feeder, column feeder and bottom feeder respectively and are used for composite fish farming.

Now-a-days in India polyculture is progressing well in larger ponds on commercial scale. In W. Bengal at Anjana fish Farm, 4,000 kg/hectare/year production of fish has been recorded. In Central Fisheries Research Institute investigations are going on for best combination of fishes for polyculture in Indian water sources. It is hoped that successful steps of polyculture would certainly increase the fish production rate in our country.

Non-Conventional Methods of Fish Culture

Culture of fish and shell fishes in pens and cages can not only help in boosting the fish production of the state but also it can helps in uplifting the economic status of the poor people of the state.

1. Pen Culture: Pens are enclosures, a blocking device which acts as a barrier preventing the entry of undesirable animals and fishes ensuring the safety of the cultured stock. The system ensures higher survival rate and better yield. The farming system can be operated in the marginal areas of the large water bodies and could be considered as an eco-friendly type as it does not interfere with other activities or pollute the environment. The pens are constructed by split bamboo and interlacing them with either can strips, coconut or nylon twines. Bamboo screens thus constructed are erected in an ideal site (where ingress and egress of water is not extreme) over the framework of bamboo poles where pen screens are strongly fixed. The screens except the shore side cordon the entire area under the pen operation.

2. Cage culture: The culture of fish in cages has shown the possibilities production of carp seeds and production of table fish through monoculture of high priced air breathing fishes like- magur, singhi, etc. and freshwater prawns.

Some of the advantages of cage culture of fishes are:
- It provides private ownership in public waters.
- In this type of fish culture control of competitor species of fishes and also predators is easy.
- Cost of construction of cage is less.
- Initial expenditure of this fish culture practice is less.
- High yield of fish.
- Good economic return.
- Ornamental Fish Breeding

Keeping colourful and fancy fishes known as ornamental fishes, aquarium fishes, or live jewels is one of the oldest and most popular hobbies in the world. The growing interest in aquarium fishes has resulted in steady increase in aquarium fish trade globally.

Technology At present in India, hundreds of exotic and indigenous ornamental fish varieties are being bred under captive condition. Majority of the production goes to domestic market and to some extent for export. A generalised production cycle of ornamental fishes is given below. There are quite a large number of tropical aquarium fishes known to the aquarists. While many of the fishes are easy to breed, some of these are rare, difficult to breed and expensive. Most of the exotic species can be bred and reared easily since the technology is simple and well developed. It is advisable to start with common, attractive, easily bred and less expensive species before attempting the more challenging ones.

Haryana Fisheries Department has established an Ornamental Fish Hatchery a Fish Seed Farm Saidpura, Karnal. The department provide technical known for rearing and breeding of the ornamental fishes so that unemployed youths, fish farmers can adopt the latest technology to enhance fish production and simultaneously raise their income.

Culture/rearing: The culture/rearing of these fishes can be taken up normally in cement tanks. Cement tanks are easy to maintain and durable. One species can be stocked in one tank. However, in case of compatible species two or three species can occupy the same tank. Ground water from dug wells / deep tube wells/ borewells are the best for rearing fish. The fishes reach marketable size in around 4 to 6 months. Eight to ten crops can be taken in a year.

Feeding: Young fish are fed mainly with Infusoria, Artemia, Daphnia, Mosquito larvae Tubifex and Blood worms. For rearing, formulated artificial or prepared feed can be used. At present no indigenous prepared feed for aquarium fish is available. The amount and type of food to be given depends on the size of the fry. Feeding is generally done twice in a day or according to the requirement. For rearing from fry stage dry/ prepared feed can be used.

Breeding: Ninety five per cent of our ornamental fishe xport is based on wild collection. Such capture based export is not sustainable and it is a matter of concern for the industry. In order to sustain the growth it is absolutely necessary to shift the focus from capture to culture based development. Moreover, most of the fish species grown for their ornamental importance can be bred in India successfully. Organised trade in ornamental fish depends on assured and adequate supply, which is possible only through mass breeding.

The method of breeding is based on the family characteristics of the fish. The success of breeding depend on the compatibility of pairs, the identification of breeders which is a skill gained through experience. Generally the brooders are selected from the standing crop or purchased and reared separately by feeding them with good live food. However, it is always better to buy good brood stock and replace the breeders. Otherwise, the original characteristic of the species keeps on getting diluted because of continuous inbreeding. Brooders especially egg layers should be discarded after few spawnings.

Health care: Water exchange, is a must for maintaining water quality conducive for the fish health. Only healthy fish can withstand the effects of transportation and fetch a good price. Permitted chemicals / antibiotics, vitamins, etc. can also be used for preventing / treating diseases.

Market: At present the market is mainly domestic and the demand is increasing steadily. The export market for indigenously bred exotic species is also fast growing and encouraging.

IntegratedFishFarming

Fisheries Department provides technical and financial assistance from integrated fish farming. The integrated fish farming practices utilize the waste form different components of the system viz. livestock, poultry, duckery, piggery and agriculture byproducts for fish production. 40-50 kg of organic wastes are converted into one kg of fish, while the pond silt is utilized as fertilizers for the fodder corps, which in turn is used to raise livestock. The system of integrated farming is very wide.

The system provides meal, milk, eggs, fruits, vegetables, mushroom, fodderand grains in addition to fish. It utilizes the pond dykes which otherwise remain utilized for the production of additional food and income to the farmer. The possible integrated farming systems are given below:-

a) Fish cum Agriculture System	b) Fish cum Animal System
Fish cum Paddy Culture	Fish cum Diary
Fish cum water chestnut	Fish cum Pig Farming
Fish cum Pappaya	Fish cum Rabbit Farming
Fish cum Mulberry	Fish cum Poultry
Fish cum Mushroom	Fish cum Duck Farming

Lacustrine fisheries

The culture of fish in takes constitute lacustrine fisheries. Natural lakes of 0.72 million hectares and man made lakes of 65 million hectares are available for fish culture in India.

Sewage fisheries

In many countries fishes are introduced and cultured on commercial basis in sewage canals and ponds. The sewage is used as fertilizer in culture ponds and as feed for fish.

Exotic Fishes

The fishes imported into a country for fish culture are called as 'Exotic fishes' and such fish culture is known as 'Exotic fish culture'. A large number of exotic food fishes have been imported and a few of them have achieved good success for fish culture industry in India.

1. The Gourami *(Osphronemus goramy)* isnative of Indonesia and has been introduced in Calcutta. In 1809 same fish was introduced in the ponds of Madras and Nilgiris from Mauritius.
2. The Tench *(Tinca tinca)* was imported from England and introduced into the Ootacamund lake. It was quite successful fish for fish culture.
3. The Crucian carp *(Carassius carassius)* is a golden carp, native of central Europe. It was introduced in 1870 in Ootacamund lake and further transferred in ponds of Nilgiris.
4. The common carp *(Cyprinus carpio)* is native of China and was introduced in Shri Lanka from Prussia in 1914.

There are three varieties of this carp *viz.,* Mirror carp (var. *specularies),* Scale carp (var. *communis)* and Leather carp (var. *nudus).* The mirror carp is most common. In last decade the mirror carps were introduced into plains of Nainital district, Ram Nagar, Kichha as well as Mirzapur district and some other places in U.P. where it has adapted well and is surviving. Other exotic fishes*viz., Tilapia mussambica* from Bangkok, *Punctius javanicus* from Indonesia *and Hypopthalmichthys molitrix* from Japan have alsobeen imported and are among good food fishes.

The introduction of exotic fishes in fish culture does not appear to harm the indigenous fauna of fishes but some workers have warned that introduction of exotic fishes is not very safe as they may upset the balance of natural fauna. Strict quarantine restrictions should be exercised during import of fishes.

Use of Inorganic and Organic Nutrients by Fish

In systematic fish culture it is usual to add supplementary food to obtain higher yields, as the water may support only a limited crop of zooplankton even after fertilization. Additional advantages that accrue from supplementary feeding are (i) a higher density of stocking for better use of available natural food, (ii) better growth on a mixed diet, and (iii) residual manurial value of the feed. The nutrients in the raw or treated sewage can be classified broadly as inorganic and organic nutrients. The inorganic nutrients are nitrogen, phosphorus and potash. These are also produced by the mineralization is accelerated in the oxidation ponds used for sewage water treatment. Detergents, which find their way in domestic sewage are also a source of phosphorus. The above organic nutrients are taken up by algae, which grow in abundance in sewage water. The other requirement of the algae to grow is sunlight and warm temperature which are in abundant supply in India. A simple chemical model of primary production in approximately stoichiometric proportion is provided by:

$$122CO_2 + 16NH_4 + PO_4 + 58H_2O \xrightarrow{Light} C_{122}H_{179}O_{44}N_{16}P + 131O_2 + H + \text{algal biomass}$$

The carbon is obtained from CO_2 produced from decomposition or from the carbonate and bicarbonate reserves in water. It is favorable for fish that free carbon dioxide (by free we mean it is not combinedwith anything) levels rarely exceed 20 mg/L (milligrams per liter), because most fish are able to tolerate this carbon dioxide level without bad effects. When several days of heavy cloud cover occur, plants ability to photosynthesize is reduced. When that happens in a pond containing lots of plant life, fish can be hurt in two ways: by low dissolved oxygen and by high carbon dioxide levels. Carbon dioxide quickly combines in water to form carbonic acid, a weak acid. The presence of carbonic acid in water ways may be good or bad depending on the water's pH and alkalinity. If this water is alkaline (high pH), the carbonic acid will act to neutralize it. But if the water is already quite acidic (low pH), the carbonic acid will only make things worse by making it even more acidic. Phenols are organic chemicals produced when coal and wood are distilled and when oil is refined. Phenols are found in a number of products from organic wastes to sheep dip. Although phenols are very toxic, dilute solutions of phenol (carbolic acid) are used as disinfectant. Algal cell contains about 50 to 60% protein, 15 to 20% carbohydrate, 10 to 15% fat and 15% minerals, which renders it to be a very valuable nutrient. The phyto and zooplankton growing in sewage water form a natural food of fishes. The algal cells are organic nutrients converted from the inorganic nutrients. The yield of algae may vary from 2000 kg per hectare per month in winter to 13000 kg per hectare per month in summer, depending upon other conditions (Datta, 2000). Some species of fish may also utilize the organic matter in sewage by directly eating the fine particles of organic matter. Blue green algae grow on the organic material in sewage. Filter feeding fish eat planktonic algae, weed eating fish eat macrophytic growth, detrivorous fish feed in the mud and benthic feeders eat snails and shrimps.

Fish Culture and Water Pollution

Water pollution is a serious event in many countries of the world where, along with the industrialization programmes, population pressure increase also takes place much faster than the treatment facilities. Effluents of fertilizer factories, pulp mills, textile and paper mills, distilleries, nuclear reactors, thermal plants and use of pesticides, herbicides, etc. change the quality of water resources with regard to their chemical, physical and biological parameters. The fishes are directly influenced due to these toxic chemicals which cause mortality and several diseases as a result the growth are ceased. The breeding capacity of fishes is also hampered due to these toxicants. Pollution affects the fish culture programme directly and indirectly both as given:

1. Change in physical, chemical and ecological nature of water resources.
2. Decreased percentage of dissolved oxygen in water affects the respiratory function of gills.
3. Cause reduction in phyto and zooplanktons.
4. Destroy the spawning grounds and other nourishing materials.
5. Fish flesh is loaded with pollutants.
6. Flesh becomes unpalatable.
7. Sewage and silt cause quick destruction of Ponds.

Therefore, water resources should be kept free from these toxic pollutants which would be of great help for fish industry and human beings also.

Law and regulations

At the central level, several key laws and regulations are relevant to fisheries and aquaculture. These include the British-era Indian Fisheries Act (1897), which penalizes the killing of fish by poisoning water and by using explosives; the Environment (Protection) Act (1986), being an umbrella act containing provisions for all environment related issues affecting fisheries and aquaculture industry in India. India also has enacted the Water (Prevention and Control of Pollution) Act (1974) and the Wild Life Protection Act (1972). All these legislations must be read in conjunction with one another, and with the local laws of a specific state, to gain a full picture of the law and regulations that are applicable to fisheries and aquaculture in India.

Research and training

Fisheries research and training institutions are supported by central and state governments that deserve much of the credit for the expansion and improvements in the Indian fishing industry. The principal fisheries research institutions, all of which operate under the Indian Council of Agricultural Research, are the Central Marine Fisheries Research Institute at Kochi (formerly Cochin), Kerala; the Central Inland Fisheries Institute at Barrackpore, West Bengal; and the Central Institute of Fisheries Technology at Willingdon Island near Kochi. Most fishery training is provided by the Central Institute for Fishery Education in Mumbai, which has ancillary institutions in Barrackpore, Agra (Uttar Pradesh), and Hyderabad (Andhra Pradesh). The Central Fisheries Corporation in Calcutta is instrumental in bringing about improvements in fishing methods, ice production, processing, storing, marketing, and constructing and repairing fishing vessels. Operating under a 1972 law, the Marine Products Export Development Authority (MPEDA), headquartered in Kochi, has made several market surveys abroad and has been instrumental in introducing and enforcing hygiene standards that have gained for Indian fishery export products a reputation for cleanliness and quality.

Programmes

The Government of India launched National Fisheries Development Board in 2006. Its headquarters are in Hyderabad, located in a fish shaped building. Its activity focus areas are:

- Intensive Aquaculture in Ponds and Tanks
- Fisheries Development in Reservoirs.
- Coastal Aquaculture
- Mariculture
- Sea weed Cultivation
- Infrastructure: Fishing Harbours and Landing Centres
- Fish Dressing Centres and Solar Drying of Fish
- Domestic Marketing
- Technology Upgradation
- Deep Sea Fishing and Tuna Processing

Management of Fisheries

The term 'fisheries' refers to the capture of aquatic animals for the use of human beings. Fisheries management includes all the skillful steps taken by man for the complete exploitation of aquatic resources. Fishery is the aquatic counterpart of agriculture. Both fisheries and agriculture are expected to step up the production of food. While agriculture is progressing towards green revolution and fishery is progressing towards blue revolution.

An intensive system of fisheries requires skillful management. According to the FAO, there are "no clear and generally accepted definitions of fisheries management". However, the working definition used by the FAO and much cited elsewhere is:

> *The integrated process of information gathering, analysis, planning, consultation, decision-making, allocation of resources and formulation and implementation, with enforcement as necessary, of regulations or rules which govern fisheries activities in order to ensure the continued productivity of the resources and the accomplishment of other fisheries objectives.*

The following are the fish management procedures:

1. Selection of species

The species selected should have the following characteristics:

- It must be marketable
- It must be pleasant to taste
- It must have successful and conducive breeding habits
- It must be easily cultivable
- It should be adaptable to crowding

2. Site selection

Topography should be as flat as possible, water should be freely flowing, but flooding should be avoided. The place should be easy for inspections, maintenance and repairs.

3. Water

A constant supply of high quality water should be available. Water should be fertile. The fertility of water is determined by the presence of salts like CO_3, NO_3, SO_4, PO_4 etc. these salts accelerate the growth of phytoplankton which forms the food for fishes.

4. Soil

The soil should be fertile and it should retain water. The power of the soil to retain water is a must. Clayey soil is more suitable.

5. Culture Tanks

Intensive fish culture requires four types of culture tanks. They are the hatchery tank, the nursery tank, the rearing tank, and the stocking tank. Hatchery tanks are shallow tanks used for storing eggs and larvae for 2 to 3 days. Nursery tanks are used to maintain tiny fry for about 15 days. The rearing tank is used to rear fingerlings for 2 to 4 months. Stocking tanks are the largest tanks for rearing fishes up to the marketable stage.

6. Preparation of Culture tanks

During summer, the ponds must be desilted. They must be allowed to dry in the sun for a few days. The unwanted vegetation should be plucked out. Desirable vegetation should be planted. Inlets and the outlets should be provided with small wire-meshed shutters.

7. Liming the pond

Ponds should be treated with lime in the form of ground lime stone ($CaCO_3$), slacked lime ($Ca(OH)_2$) and quick lime (CaO) for the following purposes:
   (i) To correct the acidity of soil and water
   (ii) To establish strong pH buffer system
   (iii) To speed up the decomposition of organic matter
   (iv) To act as a disinfectant against the diseases in fishes.
   (v) To supply one of the essential nutrients, namely calcium.

8. Pond fertilization

Ponds should be properly manured by either organic manure (cowdung, oil cakes etc) or by inorganic manure or by both. Manure helps the flourishing of phytoplankton which forms the base of the food chain of fishes.

9. Predator control

Predatory fishes should be eradicated to avoid the mortality of fry and fingerlings. Predators can be killed by the application of mahua oil cake. It is an effective fish poison cum fertilizer. The oil cake is applied at the rate of 2000 to 2500 kg per hectare. The poisonous effect of the oil cake may persist for about 15 days.

10. Weed control

Too much of weeds is destructive. They occupy more space and cause O_2 depletion. During summer the decomposition of accumulated vegetation caused biological oxygen demand (BOD). Weeds can be controlled by weed cutting machines, by biological control or by herbicides.

11. Disease control

Fisheries are much affected by parasites and diseases. Diseases are caused by viruses, bacteria, fungi, protozoans, nematodes, leeches etc. most of the diseases can be treated by dipping the fish in $KMnO_4$ solution.

12. Shelter and Cover

Culture tanks should be provided with provisions for shelter and cover. They enhance the survival of fish in a number of ways. They provide shades and help the fish to escape from the predators.

13. Habitat improvement

High rate of production and rich harvests are achieved in fish production by improving the carrying capacity of the pond. The carrying capacity of the pond depends on food supply, shelter, cover, unpolluted water, abundance of breeding sites, proper oxygen levels, suitable water-temperature etc.

14. Supplementary feeding

The rate of production is increased by supplementary feeding. The common supplementary feeds are groundnut cake, mustard cake, rice bran, wheat bran etc. They can be kept in trays suspended in the corners of ponds. The supplementary feeds are consumed directly, and anything left overs serve as manure that helps the growth of plankton.

15. Fish sampling

Harvesting should be done after the fish attained maximum weight. After harvest, fishes are marketed as such or processed for storage. During harvest, the growth of the fish in length and weight is recorded. The average rate of production is 3000 to 5000 kg per hectare per year in polyculture.

16. Harvesting

Sampling can be done once in a month to assess the growth of fishes. During each time of handling, the fish should be given dip treatment in $KMnO_4$ for 2 minutes for disinfection.

17. Introduction

Introduction is the stocking of fishes in a new geographical region. It is of two types, namely native introduction, and exotic introduction. Native introduction is the stocking of fishes of the same country. Exotic introduction is the stocking of fishes of other countries.

18. Restriction

Fishing should be prevented in certain seasons. The seasons during which fishes are in breed are called closed seasons. During the closed seasons there should be no fishing. The size-limit of the fish should also be taken into account. Fishermen should not be allowed to catch young fishes.

19. Artificial spawning sites

Certain fishes cannot breed in ponds. But they will breed in running water. For such fishes, artificial running water devices should be provided. Sand and gravel must be placed at the bottom. To help migratory fishes to reach their spawning ground, fish ladder's should be placed in big dams.

20. Selection of superior fish

New varieties of breeds should be invented for better growth and disease resistance. This is done by genetic improvement. Genetic improvement is brought about by fish hybridization, introgression, polyploidy, gynaegenesis and selection.

21. Seed selection

Fish seeds includes fish eggs, fish fry and fingerlings. Procuring fish eggs, fish fry and fingerlings for intensive fish culture is called seed collection. Fish seeds can be obtained from two sources, namely wild collection and hypophysation.

A. Wild collection: The collection of seeds from natural habitats such as rivers, ponds, lakes, estuaries etc. is called wild collection. Wild collection has certain disadvantages such as

- The seeds of predator fishes also happen to be collected.
- The seeds can be collected only during breeding seasons.
- There will be heavy mortality during collection and transport.
- There is likely to be collection of heterogeneous seeds.
- There is no scope for stock improvement.

B. **Hypophysation:** It is a technique by which females are induced to spawn by the injection of pituitary glands.

By this method pure seeds can be produced on a large scale at any time. Hypophysation has the following advantages:

- Seeds can be produced on a large scale.
- Pure seeds can be produced
- Seeds can be obtained when required.
- Seeds are free from predatory seeds.
- Stock improvement by hybridization and polyploidy is possible.
- The chance of fertilization is high.

Importance of fishery

Fishery has the following contributions to human welfare:

- Next to agriculture, fishery is the most important source of food.
- Fish is rich in proteins, minerals and vitamins.
- Fish contains the cheapest animal protein. Hence it is aptly described as 'poorman's food'.
- Fish contains less fat.
- Fish is easily digestible.
- Fish liver oils contain a large amount of vitamin A and D.
- Fish is an important component of poultry and cattle feed.
- Fish glue, fish skin, fish fin and isinglass are important by by-products of fish, useful to man.
- Fishes help in biological control of diseases.
- Fishing is an important hobby.
- Fishes kept in homes and public aquaria provide enjoyment and education.
- Fisheries provide employment opportunity for thousands of people.

SHEEP – WOOL INDUSTRY

Classification

Kingdom: Animalia; Phylum: Chordata; Class: Mammalia; Order: Artiodactyla; Family: Bovidae; Genus:*Ovis*; Species:*aries*

Introduction

The history of the sheep industry began in Central Asia 10,000 years ago as early as then, man had discovered that sheep could provide two of life's essentials-a soft warm covering and food. The use of wool as a textile fibre dates back to 4000 B.C. when it was used as such by the Babylonians. Even through England tried to discourage the wool industry in the American colonies, yarn and even sheep were smuggled into the new country.

For many centuries people have exploited the characteristics of the wool fibre. Most know the value of woollen clothing for keeping warm and that it retains this feature even when wet.In the home woollen blankets, carpets and materials all make life more comfortable.Sheep skin rugs are widely used on beds for long term patients to help prevent bed sores, and sheep skin car seat covers are very popular.

In India Woolen textiles and clothing industry is relatively small compared to the cotton and man made fibre based textiles and clothing industry. However, the woolen sector plays an important role in linking the rural economy with the manufacturing industry, represented by small, medium and large scale units. The product portfolio is equally divergent from textile intermediaries to finished textiles, garments, knitwears, blankets, carpets and an incipient presence in technical textiles. Wool industry is a rural based export oriented industry and caters to civil and defence requirements for warmer clothing. India has the 3rd largest sheep population country in the world having 6.40 crores sheep producing 43.30 million kg of raw wool. Out of this about 85% is carpet grade , 5% apparel grade and remaining 10% coarser grade for making rough Kambals etc. Average annual yield per sheep in India is 0.9 Kg. against the world average of 2.4 Kg. A small quantity of specialty fibre is obtained from Pashmina goats and Angora rabbits. The domestic production of wool is not adequate, therefore, the industry is dependent on imported raw material and wool is the only natural fibre in which the country is deficient. The woolen industry in the country is of the size of Rs. 10,000 Crore and broadly divided and scattered between the organized and decentralized sectors. The organized sector consists of: Composite mills, Combing units, Worsted and Non

Worsted spinning units, Kintwears and Woven Garments units and Machine Made Carpets manufacturing units. The Decentralize Sector includes Hosiery and knitting, Power-looms, Hand knitted carpets, Druggets, Namadahs and Independent dyeing, Process houses and Woolen Handloom Sector. There are around 958 woolen units in the country, majority of which are in the small scale sector. The industry has the potential to generate employment in far-flung and diverse regions and at present provides employment in the organised wool sector to about 12 lakh persons, with an additional 12 lakh persons associated in the sheep rearing and farming sector. Further, there are 3.2 lakh weavers in the carpet sector.

During the XIth Five Year Plan period, the Government implemented Integrated Wool Improvement & Development Programme (IWIDP), Quality Processing of Wool and Woollen Products and Social Security Scheme for sheep breeders in the country for the growth and development of the wool and woollen industry. The scheme was implemented by the Central Wool and Woollen industry.

Although India is among the leading countries in terms of sheep population, the wool productivity is much lower than the world average. Also, given the inadequate quality and quantity of wool produced in India, the country imports substantial amount of wool. The wool growing community as also the various user industries are currently facing several challenges on account of the current state of affairs, and adequate policy intervention is required to improve the prospects of the over all industry. This report attempts to discuss the international scenario with respect to the wool industry; the Indian scenario; D and B India's outlook on the prospects of the Indian Wool and Woollen products Industry; the key issues and concerns faced by the industry currently, and proposed solutions to meet the various challenges facing the industry.

International Scenario: World production of greasy wool has been on a decline since the past several years, with production having fallen from 3.39 million tonns in 1990 to 2.11 million tonns by 2008. Australia, China and New Zealand are the world's leading producers of wool. Australia, the largest producer of wool in the world, has been witnessing falling production over the years, mainly on account of fall in sheep population. On the other hand, production of wool in China, the second largest wool producer, is on an upward trend. In line with the fall in global demand for woolen products, the year 2008 witnessed significant decline in exports across different categories.

Indian Scenario: India is the seventh largest producer of wool and contributes 1.8% to total world production. In sheep population, India ranks among the leading five countries in the world. However, at 0.8 kg/sheep/year, wool productivity in India is much lower than the world average of 2.4 kg/sheep/year.

Bulk of the wool produced in India is of coarse quality and used mainly in the manufacture of hand-knitted carpets. Demand for wool exceeds domestic production, since India produces a lot of value added products that are exported. Thus, substantial amount of raw wool is imported.

Australia, New Zealand, Pakistan, China etc. are the major countries from which India meets its import requirements for raw wool. On the export front, UK, Italy, USA, Dominican Republic and UAE are the major countries for India's exports of woolen yarn, fabrics and made-ups, while USA, UAE, UK, Germany and France are the leading markets for India's exports of readymade wool garments. During 2008-09, overall exports of wool and wool blended products are estimated to have declined by 8.4% to around Rs 5,064.3 crore.

The wool yielding sheeps are inhabiting the arid regions of Northern India specially in plains as well as in the hills. The important places *viz.,* Saurashtra, North Gujarat, Kutch, Kashmir and the foot hill districts of Himachal Pradesh and Garhwal are most favourable for providing the natural conditions suitable for raising fine woolen types of sheep. The largest sheep population has been recorded at plateau of Deccan and Vindhya mountains. The wool obtained from the sheep of Kashmir is finer in comparison to that of other places. The Magra and Chokla are the best breeds from Bikaner and Kutchi from Joria (Rajasthan) which are famous for the superior type of carpet wools.

CLASSIFICATION OF WOOL

Wool yield and grade are the two main factors that are given consideration when buying and selling wool. As a producer, knowing the quality of the wool that you produce will help when making management decisions.

Wool Classification – is the categorization of wool (i.e. Bright, Semi-Bright or Dark) by subjective measurement of the amount of clean wool in a given fleece. Manufacturers buy wool on a clean yield basis (where all foreign matter and grease has been removed). The presence of foreign matter impacts both the quality and processing costs of manufactured wool products. The inability to remove foreign matter is reflected in the quality of the end product. Increased processing costs are due to the need for extra processes to remove foreign material and maintenance of machinery damaged by foreign material. The breed of sheep and the general care of them, as well as geographic and climatic conditions can all affect wool class.

Wool Grade – is determined by texture (crimp style), length and diameter of the fibre. Different breeds of sheep produce different grades of wool which in turn have different uses. Range Wool is from the breeds of sheep producing the finer grades of wool. Range wool is heavier with natural grease.

Fine – 22/23 Micron Wool, 2.5 to 3 inch Staple Length
Half – 23/24 Micron Wool, 2.5" to 3.5 inch Staple Length
Range 3/8 – 26/27 Micron Wool, 3 to 3.5 inch Staple Length
Range 1/4 – 30/31 Micron Wool, 3 to 4 inch Staple Length

CHARACTERISTICS OF WOOL

Its unique physical and chemical characteristics have been responsible for its great versatility and high value in the manufacture of clothing. Although many scientists have tried, they have not been able to produce a synthetic fibre with the same specific characteristics as wool.

1. Fineness of wool

The fineness, or thickness, of the fibre is the most important single characteristic of wool, greatly influencing its economic value. The degree of thickness determines whether the finished fabric will be a fine dress material or a coarse floor covering. In the wool trade, fineness is either judged visually or measured precisely – it is on this basis that the grades of wool are determined.

Increased emphasis on an exact and highly descriptive method of describing wool grade has produced a measuring system in which individual fibres are accurately measured. The unit of measure is the micron, which is one millionth of a metre or 1/25,000 of an inch. Fineness is expressed as the mean fibre diameter. From a casual observation it would appear the fibres growing on a sheep's skin are relatively uniform in thickness. However, the fibre thickness may vary from 10-70 microns within the same fleece . Rambouillet fleeces usually average 20-25 microns in fibre thickness, whereas Lincoln fleeces average 35-40 micron.

2. Length of fibre

Good length of fibre is essential for the production of a superior worsted yarn. Length of fibre is determined to a large extent by the breed of sheep; that is, it is largely an inherited factor, but it can be influenced by nutrition. Experiments have shown that a high plane of nutrition will increase the fibre length by as much as 170% of that produced on a low plane of nutrition. For maximum production the animal must be well fed.

The following minimum, unstretched lengths are required for the various grades of wool before they can be classed as "staple wool."

Fine staple 5.0 cm
One-half staple 6.5 cm
Three-eighths staple 7.5 cm
One-quarter staple 7.5 cm
Low one-quarter staple 10.0 cm

3. Strength of fibre

To withstand the stress of manufacture and produce a strong, long-wearing fabric, wool must possess tensilestrength. To be classed as a "strong wool," a high percentage of fibres must pass through the carding, combing and spinning processes without breaking.

Wool produced under normal range conditions, where the sheep have received sufficient feed, usually has adequate strength. However, there are two conditions that may cause a lack of strength.

One condition is known as "tender wool," i.e., fibre that is weak throughout its entire length. This is usually due to the sheep having some chronic disorder, being on a low plane of nutrition for an extended period, or being old.

The second condition is a break, or definite weak spot, at a particular location on the fibre. This is noted readily when the wool is stretched, as it breaks squarely across the staple. Sudden illness, starvation during a bad storm, or overfeeding of concentrates, are mainly responsible for this condition. There can also be some difficulty experienced with a fleece break at lambing time. For this reason, it has become common practice to shear as soon as possible before or after lambing so that shearing will occur at the break; thus the effect of the break will not be apparent in the fleece.

4. Crimp

Crimp is the term used to designate the natural waviness of wool fibres. The number of crimps will vary from 1 to 30/2.5 cm, depending on the degree of coarseness. More crimps are present in the finer wools. Well crimped wool will spin more easily and produce a finer and stronger yarn with less wastage than poorly crimped wool. Uniformity of crimp is associated with uniformity of fineness and length, and is a sign of superior quality.

5. Colour

The normal colour of wool from the improved breeds of sheep is white, but a small percentage of it may be brown, black, or grey. Generally, manufacturers demand that the wools used in processing be scoured out completely white to ensure that the future colour of the fabric will not be affected by the natural colour of the fibres.

The presence of dark or off-colour fibres in white fabrics causes them to dye unevenly and in addition, makes them unsuitable for pastel colouring. The black-faced breeds, for example Suffolk and Hampshire, tend to have black or brown fibres mixed with the white portion of the fleece on their legs and head and occasionally throughout the main portion of the fleece.

6. Felting properties

The capacity to felt, one of the characteristics peculiar to wool and only a few other hair fibres, is attributed primarily to the presence of scales on the surface of the fibre and to its crimpy nature. Under the influence of heat, moisture, alkali and pressure, the fibres form a wool pad or cloth, that can be used for wearing apparel. Common items illustrating this type of manufacture are felt hats, felt boots, felt socks and felt cloth. Woven goods may also be subjected to manipulation and pressure in hot, soapy water to produce a felt surface. This process of finishing cloth, known as felting, is commonly employed in the manufacture of melton and billiard cloth.

7. Elasticity

Elasticity is the ability of wool to return to its original form after having been forced out of shape by pressure. This is one of the peculiar characteristics of wool that makes it superior to other textile fibres. Yarn from highly elastic wool can withstand the stress of manufacture more readily, and the garments produced will hold their shape better than those produced from wool lacking this property. In general, fine wools are more elastic than coarse wools.

8. Yield and shrinkage

Yield is the amount of clean wool that remains after scouring, expressed as a percentage of the original grease weight. For example, a 4.50-kg grease weight fleece producing 2.25 kg of clean wool has a yield of 50%. In other words, yield represents that portion of the raw fleece available for manufacturing purposes. Shrinkage is the weight that wool loses when scoured, expressed as a percentage of the original grease weight. Shrinkage results mainly from the removal of dirt, manure, seeds, burrs, chaff, straw, sweat salts and wool grease. Because wool processors are interested only in the quantity of clean wool present in the clips they buy, they are able to pay proportionately more for the lighter-shrinking wools.

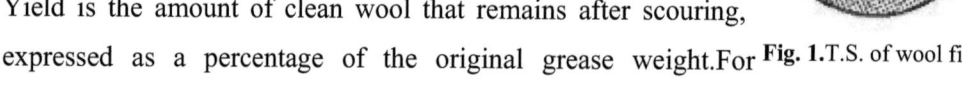

Fig. 1. T.S. of wool fiber.

CHEMISTRY OF WOOL

1. Physical Properties

The colour of the wool varies from species to species of the sheep and also the climatic conditions of the area. The real wool fibre is hygroscopic, elastic, durable, bad conductor of heat and is not easily inflammable. Due to its nature (bad conductor of heat) more heat is produced when the wool is wet. The microscopic study of the wool fiber seems to have cellular structure. In the transverse section of the wool (fig. 1) fiber two regions are distinguishable i.e., *a* central core of hard cells in the periphery and a medulla some-what softer, situated in the centre. The affinity of wool to dye absorption and the easy twist are the characteristic features of pure wool. The diameter of the wool fiber has been noticed' to be 12 to 80 micron. The wool produced in India has a remarkable property of regaining the original shape when pulled and has abrasive resistance.

2. Chemical Properties

The wool fiber is made up of keratins which are actually, the polymers of the proteins and have higher sulphur contents. It consists of a number of polypeptide chains of amino acids. The various amino acids which constitute the wool protein are: arginine, histidine, lysine, alanine, methionine, threonine, tyrosine, cystine, leucine, iso-leucine and valine.

STRUCTURE OF WOOL

Wool in its simplest terms is a fibrous protein. Proteins are polymeric substances with relative molecular masses of many thousands. The building blocks of proteins are about twenty aminoacids all but one of which have the formula $^+NH_3$-CHR-CO_2^-, and all of which have the same stereochemistry around the chiral carbon atom as shown. (Note that two aminoacids have additional chiral carbons in their side chains, R.) When the amino group of one molecule condenses with the carboxylic acid group of second molecule to form an amide, (or peptide) link, a dipeptide is formed, $^+NH_3$-CHR-CO-NH-CHR-CO_2^-. Condensation with a further aminoacid gives a tripeptide, and the process continues to form a polypeptide. With twenty different R groups the polypeptide can be likened to a string of coloured beads, each different coloured bead representing an aminoacid "residue" with a different R group. The nature and position of the R groups give the protein its unique properties.

Wool in reality is a complex biological structure, composed of an assemblage of cells of differing types held together by modified cell membranes and containing proteins of three major types. Although in much early wool chemistry, for reasons of simplicity, the biological structure and diverse composition of the fibre was overlooked, modern wool chemists must constantly relate wool chemistry to the biological processes of fibre formation. Thus wool fibre research embraces significant areas of cell biology and molecular genetics as well as classical chemistry.

Some elementary ideas of the complexity of fibre structure can be gained from Figure 1 and the micrographs of Figure 2. Obviously relating fibre properties to chemical structure must involve an understanding of chemistry and structure at a number of different levels of organisation.

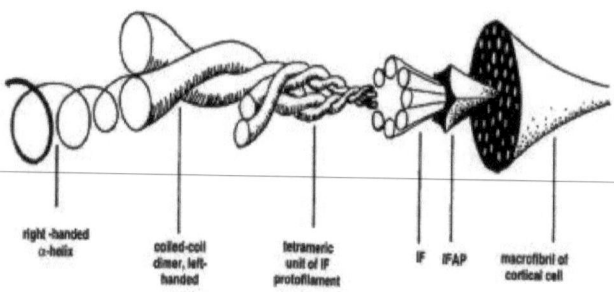

Fig. 2. Exploded diagram assembly of the teti complex. The high intermediate filament.

The protein components of the cortical cells make up the greater part of the fibre. These cells contain three main classes of protein, which may be briefly described as follows.

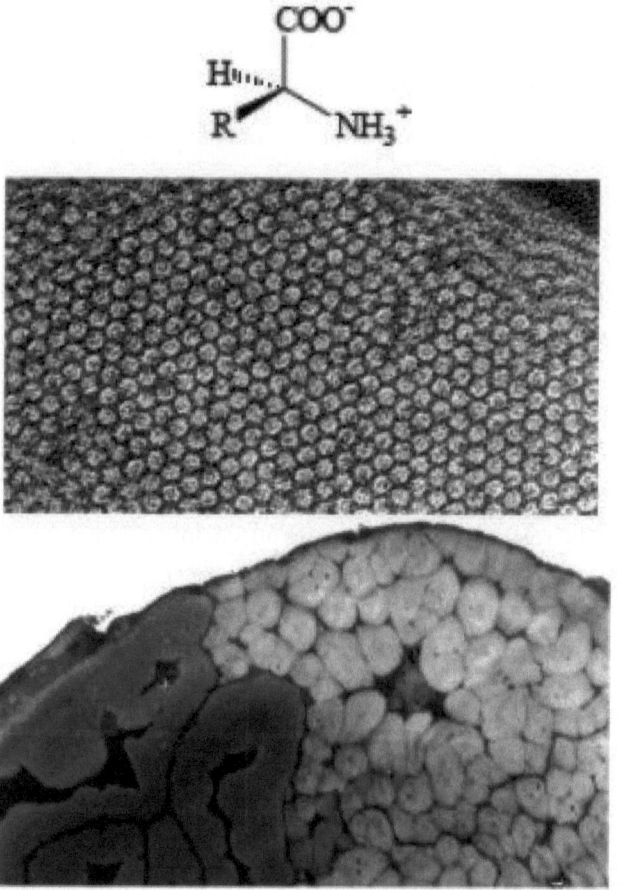

Fig. 3 (a): Electron micrographs of a transverse section of part of a cortical cell (Mesocortex) showing the closed packed-hexagonal array of intermediate filaments, sufounded by densely stained 'Matrix' of IFAP high sulphur proteins. Magnification=736000x. The separation of intermediate filaments is about 8.6 nm.

Fig. 3 (b): Electron micrograph of a transverse section of a wool fibre, showing lightly stained, differentiated macrofibrils in the orthocortex, and densely stained, fused macrofibrils in the paracortex. Note also the cell boundary structures. The dark zones, particularly in the paracortex, are regions of non-keratinous cytoplasmic remnants. Magnification=25000x. (Micrographs courtesy of J L Woods, WRONZ)

Intermediate filament (IF) proteins

These are proteins which are 'fibrillar' in that they consist largely of rod-like assemblages of α-helices, but also have end regions which are more globular or random-coil in configuration. The fibrillar IF proteins are aggregrated into complex rod-like structures about 10 nm in diameter which are known as the 'intermediate filaments' (or microfibrils). These in turn are aggregrated into much larger structures within the cell, called macrofibrils. The structure of macrofibrils varies betwen the two main types of cortical cell, ortho- and para-cortex. Figure 3 illustrates the structure of intermediate filament proteins at the simplest level of organisation, a 2-strand 'coiled-coil' of α-helices, interdispersed with non-helical linker regions.

Intermediate filament proteins typically have molecular masses in the region of 45-50 kDa (relative molecular masses of 45 000 - 50 000). Great advances in th understanding of wool and other tissues occurred in the 1980s when it was realised that the 'microfibrils' in wool and the 'intermediate filaments' of many other filamentous or membranous tissues were almost identical, and essentially the same structures. This produced an intense cross-fertilisation between wool research and biomedical research in areas such as skin and related tissues.

Intermediate-filament associated proteins (IFAPs)

This rather of clumsy piece modern jargon refers to the proteins long described in wool research as 'matrix' proteins, because they form a less-organised non-directional matrix in which the rods are embedded. (The analogy of reinforcing rods in concrete is sometimes used.) These proteins are characteristically very high in sulphur (as many as 25% of the amino-acid residues may be half-cystine), and relatively low in molecular mass (10-30 kDa). They are very numerous, and somewhat variable in relative amount depending on nutrition, etc. Matrix proteins, by virtue of their high content of cystine, are usually credited with being responsible for the disulphide bond crosslinking which gives the fibre its mechanical and chemical integrity. In fact very little is known about the conformatiom of matrix proteins, or the pattern of crosslinking involving half-cystine residues. Very recent studies of model peptides based on matrix protein fragments strongly suggest that almost all the disulphide bonds in these proteins are intramolecular, between near neighbours, to form rings closed with disulphide bonds. The long-range crosslinking is probably formed between a small number of the matrix cystine residues, and the relatively few cystines present in the IFs and their globular tails. This is consistent with inferences from applied science and technology.

High tyrosine-glycine (HGT) proteins

These proteins rich, in the amino-acids tyrosine and glycine, form a relatively minor third class of proteins. Their occurrence is restricted to certain parts of the fibre. They are relatively low in molecular mass. Their function is not clearly understood.

Helix structure

The proposal of the α-helix structure of polypeptides by the famous double Nobel-laureate Linus Pauling and his colleagues in 1951 was one of the great break throughs of protein research. In this model structure, there is a 'residue transition' (ie. the axial distance traversed by each by each aminoacid residue) of approximately 0.15 nm. This was consistent with key results from early X-ray diffraction work on filamentous α-keratin. The structure is stabilised by intra-chain hydrogen bonds involving the hydrogen on the –NH group of one residue and the oxygen of the amide bond further residues along the chain. In the classical α-helix structure, there are 3.6 amino acid residues per turn, or 18 residues in 5 turns, over an axial length of 2.7 nm (0.15 nm/residue). In natural proteins, all the aminoacids have the L-configuration and the á-helix is invariably right-handed. The mechanical association via cross-linking between the filaments and the matrix is largely responsible for the observed mechanical properties of the fibre. An on-going goal of academic wool fibe physicists is to develop a model of mechanical behaviour which accurately reproduces the observed mechanical behaviour (both in tension and in relaxation, see Figure 4).

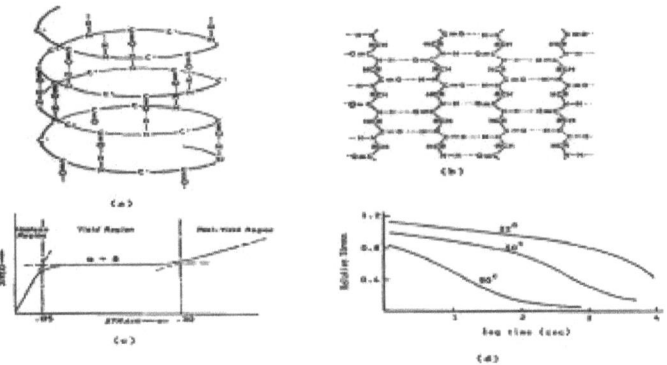

Fig. 4: Aspects of structure and mechanical properties of wool.

(a) A schematic representation of á-helix structure, stabilized by intra-molecular hydrogen bonds, as adopted by fibrillar proteins of the unstrained fibre.

(b) α-pleated sheet structure, stabilized by inter-molecular hydrogen bonds, as adopted by zones of the fibrils when the fibre is stretched to the end of the yield region.

(c) Stress-strain diagram for wool fibre, showing yield region associated with a transformation.

(d) Stress-decay curves (relative stress vs log time) for wool fibres at 40% strain, in pH 7 buffer, at various temperature (the rate of stress decay is closely related to the rate of thr thiol-disulphide interchange reaction.

Wool has a unique form of strain-stress curve, depicted in Figure 5. One of the key features is the flat yield region where the fibre extends at constant stress. It has been shown that this mechanical feature is associated with conversion of α-helical rods to extended-chain structures known as α-pleated sheets (which are also strongly hydrogen-bonded). The α conversion is an important feature of wool behaviour. At the same time as the helices are extended, the matrix is also stretched. Recovery of the fibre occurs when the stress is released, assisted by the retractive force of the strained disulphide bond network, much like a rubber. In fact, there is a lot in common between wool and the polysulphide rubbers in mechanical behaviour. Surrounding the cortical cells are protective layers of scales. It is often the case that nature has perfected a masterpiece of engineering design which man can merely copy. The design of the wool fibre is copied in high class climbing ropes which are coiled bundles which are in turn coiled, and the whole structure surrounded in a protective sheath.

WOOL SCOURING

Wool as shorn from the sheep is known as greasy (or raw) wool. Before it is suitable for further processing it must be washed to remove dirt, water soluble contaminants (called suint), and woolgrease. This process is known as scouring.

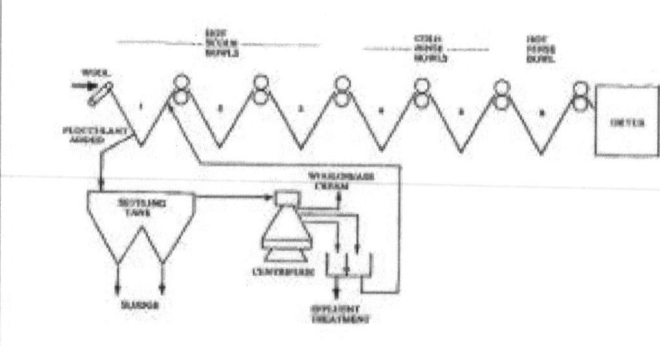

Fig. 5: Simplified layout of a typical commercial woolscouring plant showing liquor treatment loop for dirt and grease removal. The latest plants have 8 bowls, sophisticated control systems, multiple centrifuges, and liquor handling loops which are much more detailed than shown in the diagram.

The wool passes through a series of bowls, each separated from the next by large pressurized squeeze rollers. There are (usually) six bowls; the first three are scouring bowls and contain hot detergent solution, (about 60°C, 1-3 g/L); the next two bowls are cold rinse bowls; and the final bowl is a hot (60°C) rinse. The squeeze rollers minimise the carryover of contamination from one bowl to the next, and this is assisted by a general flow of the hot liquors against the direction of the wool flow. Liquor from the first (most contaminated) bowl is cycled through dirt and wool grease removal equipment, after which some, about 1 litre/kg greasy wool, is run to drain. This loss of liquor from bowl 1 is made up with liquor from bowl 2, and so on. The amount of wool a single scour train can process depends on the width of the train; the range is from 0.6-5 tonns of greasy wool per hour. The wool passes through the whole plant (scouring, drying, and baling) in about 20-30 minutes.

CONTAMINANTS OF WOOL

Raw wool contains three main contaminants, woolgrease, suint, and dirt, which combined make up some 20-30% of the fleece weight. A typical figure of the grease content of crossbred wool is about 6%.

Woolgrease

This is a very complex mixture, consisting mostly of esters of various long-chain fatty acids with long-chain alcohols and sterols. Technically it is a wax, rather than a fat, because glycerol esters are not present. The fatty acids present fall into four main classes.

- a normal paraffin series, with even carbon numbers approximately from C10 to C26.

- an iso-acid series, with the alkyl chain terminating in a (CH$_3$)2-CH- group, with even carbon numbers from C10 to C28.

- an *anteiso*-acid series, with a terminal *iso*butyl group, i. e.

$$CH_3 - CH_2 - \underset{\underset{CH_3}{|}}{CH} -$$

with odd numbers of carbons from C9 to C31.

- á-hydroxy normal and iso- acids, ie, R-CH(OH)-COOH with even carbon numbers from about C12 to C32.

The alcohols are even more complex. There are aliphatic alcohols corresponding to the same series of structures as the acids, ie

- normal alcohols from C18 to C30

- iso-alcohols from C18 to C26

- anteiso-alcohols from C17 to C27

- 1,2-diols, ie, R-CH(OH)-CH2OH, both normal and iso, C16 to C24.

However, the major part of the alcohol fraction consists of sterols; the most important member is cholesterol, followed by lanosterol and dihydrolanosterol. Smaller amounts of other sterol derivatives are present. Clearly the esters formed from these acids and alcohols form a bewildering variety, especially when one considers that the hydroxy-acids and the diols can form di-esters.

In addition to the esters, free fatty acids are present, along with other impurities. Such as soaps (metal salts of fatty acids). Calcium soaps are particularly important. These arise mainly from the fellmongery production of slipe wool with lime-sulphide depilatories, and can cause problems in the centrifugal separation of wool grease as well as in wool grease refining.

Refined wool grease is known as lanolin, which is the product resulting from deodorisation, decolorisation, neutralisation and removal of entrained water (volatiles). Lanolin is an important item of commerce. However, the most sought-after product from woolgrease is the mixed wool-wax alcohols. These are produced by hydrolysis (saponification, or soapmaking) of the esters followed by solvent extraction of the alcohols. These are widely used in toiletries and medicinal preparations because of their ability to form stable emulsions entraining large amounts of water.

Suint

Suint is the sweat of the sheep and is a complex mixture of water-soluble salts. The predominant cation is potassium; the anions include carbonate, bicarbonate, various low molecular weight mono- and di-carboxylic acids (succinic, glycollic, glutaric, etc) and smaller amounts of long chain fatty acid anions which may originate from woolgrease. Peptides and other nitrogenous substances are minor components. The long-chain suint anions are in themselves surface active and act as effective soil dispersing agents during scouring. Crossbred wools the suint pH is usually about 8.5-9.0, so the fatty acids are ionised and act as anionic surfactants (soaps) participating in the stabilisation of the woolgrease emulsion by conferring a negative charge to the emulsion particles.

Consequently, one of the key steps in emulsion destabilisation for effluent treatments is acidification, to convert those soaps to acids, thus destroying their surface charge and facilitating coalescence. This process is known as acid cracking.There is overlap between components of suint and grease scour liquors when they are analysed by the conventional solvent extraction test. Thus, some of the grease component analysed by extraction of liquor with petroleum ether is also water soluble. The distinction between suint and grease components, especially in oxidised wools, is thus poorly defined.

Dirt

Dirt consists of all the ill-defined solid fleece contaminants. It includes mineral soil, wind blown dust, vegetable matter, faecal matter (dags), skin flakes, discarded cuticle cells, and fragments of fibre broken from brittle photo-oxidised tips. In many respects, in terms of wool properties, it is the very fine mineral material, largely associated with exposed fibre tips, that is the most significant.

Detergent Selection

Until the 1960's wool was conventionally washed in an aqueous bath containing soap and alkali (usually sodium carbonate). It was necessary for soft water to be used because of the insolubility of calcium and magnesium soaps. In fact the wool textile industry developed in the Yorkshire area of the UK because of its abundance of soft water. Scouring wool with soap and alkali was a process which had to be undertakenwith some care because of the susceptibility of wool to damage by alkali. Therefore it was with some relief that the industry greeted the introduction of synthetic detergents which were unaffected by hard water and were cost effective in degreasing wool. Originally these were anionic and it was found that alkali was still needed to avoid undue loss of detergent by adsorption on to the wool. The later introduction of nonionic detergents removed this constraint, and wool was able to be washed with these detergents alone. For many years now, scouring has been carried out with nonionic detergents formed by the chain growth polymerisation reaction of ethylene oxide (EO) with an alkylphenolate ion, where the alkyl group is typically a highly branched C9 chain.

The overall reaction is:

$$R\text{-}C_6H_4\text{-}OH + n\,CH_2\text{-}CH_2\text{(O)} \xrightarrow{base} R\text{-}C_6H_4\text{-}O(CH_2CH_2O)_nH$$

although the reaction is rather more complex than implied by this equation.

Unfortunately, alcohol ethoxylate detergents which best address the needs of wool grease refiners are not the best for wool scouring. If grease properties are paramount, detergents of the general type of

$$CH_3(CH_2)_{7\text{-}9}CH_2\text{-}O(CH_2CH_2O)_6H,$$

may be used, while for best woolscouring molecules with larger hydrophobes and longer EO chains are preferred, for example

$$CH_3(CH_2)_{10\text{-}13}CH_2O(CH_2CH_2O)_9H$$

If the hydrophobe is linear, the biodegradability is also enhanced.

The detergent molecules contain a hydrophobic part which is the hydrocarbon moiety and a hydrophilic portion formed of the polyoxyethylene chain. Thus when wool grease a hydrophobic substance is removed from the fibre surface by detergent action, a stable grease micelle is formed in which the hydrocarbon moiety is embedded in the emulsified grease, while the hydrophilic polyoxyethylene chain mingles with the surrounding water. As mentioned previously anionic compounds from the wool grease and suint (fatty acid anions) also act as surface active agents and assist in stabilising the wool grease dispersion. Dirt particles (largely mineral) are also entrained in and emulsified with the grease. Soluble proteins, absorbed onto or interacting with the droplet surface, add to the complexity of the structures. The result of this process of grease emulsification (which takes place mostly at about 60^0C, above the wool grease melting temperature) is an extremely stable dispersion of wool grease and dirt which defies most conventional methods of demulsification and phase separation. This dispersion stability greatly complicates issues of effluent treatment. Dispersed particles in the liquor exist in two size ranges. The larger diameter particles (2-12 µm diameter) may be largely separated from the aqueous medium by centrifugation, leading to the recovery of wool grease as a by-product of scouring.

HARVESTING OF WOOL

To enable wool growers to achieve better wool preparation and higher financial returns the following are the recommended guidelines:

1. All sheep need to be emptied out before shearing. I.e., no feed or water to be administered to the sheep for a minimum of 12 hours prior to shearing. By carrying out this practice, the sheep's stomach and bladder will be empty and therefore the wool does not become contaminated with dung and urine. The sheep will also sit better for shearing as they do not struggle the same, which enables the shearing process to be easier for both the shearer and the sheep. Never shear wet wool or pack wet wool.

2. The belly wool needs to be kept completely separate from the fleece wool. The shearer should remove and throw the belly aside as the sheep is being shorn. Belly wool is to be packed separately.

3. All short, stained wool and tags need to be removed from the crutch area as the sheep is being shorn. This wool is kept completely separate from all other types of wool and packed separately.

4. All fleeces should be thrown onto a wool table to enable the skirting of the fleeces to be performed in a proficient manner. Chaffy or bury wool should be skirted from the fleece and packed separately.

5. The shearing board should be swept and kept clean between sheep as well as during the shearing of the sheep.

6. All fleeces should be shaken to remove any second cuts before rolling and pressing the fleeces.
7. When pressing the wool, all the different categories of wool are to be pressed separately. There should be no mixing of the different wool types during shearing, but when pressing at the end of shearing the different types of wool can be put into one bag. However, they need to be separated by sheets of newspaper.
8. All bags are to be sewd with cotton twine, do not use baling twine, wire, electric fence wire or polyprop twine to sew the wool bags.
9. All bags need to be identified as to their contents.
10. Where possible during shearing, the level of straw needs to be kept to a minimum and away from the shearing area to minimize contamination.
11. Coloured sheep are to be separated and shorn last so as not to contaminate the white wool with coloured fibres.
12. Fleece preparation incentives of up to 5 / lb is applicable for bright, high yielding fleeces that have been properly skirted and packaged (see Fig. 5).
13. Maintaining a clean shearing board and floor is an important and continuous process. It must be done before, during and after shearing to ensure a quality clip.

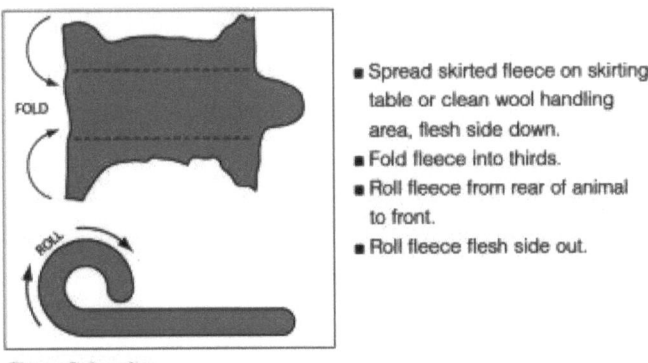

- Spread skirted fleece on skirting table or clean wool handling area, flesh side down.
- Fold fleece into thirds.
- Roll fleece from rear of animal to front.
- Roll fleece flesh side out.

Time of shearing

Years ago, most sheep in Western Canada were kept on the open ranges almost year-round, and it was the practice to shear them once a year. Shearing time was before the arrival of warm weather and after the danger of late spring storms to avoid the risk of heavy death losses. Nowadays, however, sheep with long fleece tend to become itchy in warm weather and this causes them to rub. If they roll on their backs and are unable to get up, death may result. Thus, the most suitable time for shearing is fall, winter or spring. The most critical factors in determining the time of shearing are the availability of shearers and the time of lambing. Crutching Sheep are crutched before lambing, if they are to be sheared after lambing. However, if the sheep are sheared about 4-6 weeks before lambing, the need for crutching is eliminated. Crutching involves the removal of wool from the udder, the belly area immediately in front of the udder, and between the hind legs up to the tail.

Crutching or shearing before lambing has advantages:

- Reduced danger of infection of the ewe at lambing. If difficulty occurs during lambing, assistance may be rendered much more easily.
- Reduced losses caused by bacterial infection of the digestive tract in newborn lambs sucking on sweat locks or dung tags, instead of on the teats.
- Minimized lamb losses from wool balls causing blockage in the digestive tract.
- Reduced eye soreness in nursing lambs.

Fundamentals of good shearing

Sheep producers with large flocks usually hire experienced professional sheep shearers. However, in small flocks, shearing is often done by the owner or by a neighbour who has acquired a certain amount of skill through practice. Skilled operators are essential because good shearing requires that a sheep be handled carefully and not be injured while the wool is being removed. If the shearer is experienced, the sheep will not struggle while being shorn. An unskilled shearer will have considerable difficulty in preventing the animal from struggling.

Tips for working with a custom shearer
- Book well in advance.
- Have sheep crutched beforehand.
- Pen sheep close 12 hours prior to shearing with no feed or water.
- Prepare a clean, well-lighted area with access to an electrical outlet.
- Provide plenty of head room.
- Have catch pen near the shearing area.
- Have extra help for filling the catch pen and preparing the fleece for market.

Tips for novice shearers:

- Get qualified instruction.
- Shear only dry sheep on a clean, dry surface.
- Shear belly wool first and pack separately.
- Shear coloured sheep last and pack this wool separately.
- Do not shear black face or leg fibres.
- Avoid second cuts on the wool wherever possible.

There should be no second cuts or short pieces of wool produced by cutting the staple twice. Second cuts reduce the length of fibre and, consequently, its economic value. Also, it is desirable that the fleece be removed in one piece so that it can be properly folded and rolled for market. Great care must be exercised in shearing the udders, particularly of yearling ewes; it is very easy to cut off the end of a teat and permanently damage that portion of the udder. If a sheep is seriously cut with the shears, the wound should be treated with a disinfectant and, if necessary, sewd.

Methods of shearing

Several decades ago, hand shearing was the only method available to the producer. Power shearing is today's method. It is faster than hand shearing and is easier on the sheep because it is handled for a shorter time. With trained shearers using power shears, the wool is removed with a minimum number of second cuts, thus increasing the value of the wool clips. The danger of injury caused by powershears is no greater than that caused by hand shears; sheep may be cut seriously by either method if the operators are inexperienced or careless. Researching new techniques for the past several years, research has been continuing around the world to develop a method of shearing by injecting chemical compounds into the sheep. The chemical compounds would cause, first of all, breaks in the fibre and then, a few days later, the whole fleece to peel off. Such a technique might he useful to small flock operators because they would not then have to either shear the sheep themselves or hire professional shearers. However, this method could create health and reproductive problems to the animals and make the carcasses unsafe for human consumption. Also, as shown in experiments, some chemicals do not form the breaks uniformly over the whole body within a period of time, thus causing an easy removal at some locations and difficult or no removal at other locations. There have also been problems with sunburn. It is hoped that a reliable and safe technique will soon be developed. New techniques and equipment are also being researched to reduce shearer fatigue and back injury. For example the Shear Eze machine places the sheep on a cradle that is at a comfortable height for shearing thereby minimizing the usual shearer back bending procedure.

Shearing sheds

Where large flocks are kept, it is often desirable to have a separate, permanent shearing shed. However, any building that has a waterproof roof can he used. The lambing shed is usually the most suitable building available for shearing and is one that can be converted readily for this purpose. Provision should be made within the shed for large pens to hold the sheep before shearing, a catch pen for each shearer, a smooth board shearing floor, and space for sacking and storing wool. Slatted floors are desirable in the holding pens to keep the wool as clean as possible. Through the use of these slatted floors, the sheep are raised off the ground and, as a result, have no opportunity of coming in contact with litter or manure.

Preparation for sheep shearing

- Aim for a stress-free shearing day
- Be prepared
- Have an efficient set-up

Shearing facility goals

1. Delivery of sheep to shearer with minimal effort for handler, sheep and shearer
2. Removal and preparation of wool with minimal effort – clean and carefulfleece preparation
3. Skirting table and wool packer conveniently located

Shearing facility tips

- A dry place – pens, shear floor, wool handling and storage area, all free of drips, leaks, excessive dampness
- Facilities do not need to be permanent – but arrange before shearer arrives
- Get ready the day before shearing
- Put up temporary lighting in the shearing and wool handling areas
- Shearing floor should be level to stand on
- Provide for ventilation
- Have good wiring to clipper outlet
- Sheep will be reluctant to move toward noise of shearing machine

- In chute, use a stanchioned "decoy" sheep
- If possible, have helper for moving sheep so shearer and wool handler can work without interruption
- Catch pens should hold 12-20 ewes (15 ideal) Holding pen for woolly sheep

PREPARATION OF WOOL FOR MARKET

It must be kept in mind that the manufacturer makes use of the wool only, and not of the foreign material present in the fleece. The manufacturer buys fleece wool on the basis of its clean wool content and with the exception of lanolin, everything else is waste material. Consequently, it is in the interest of the wool producer to keep waste material to a minimum by all possible, practical means. Careful preparation of the fleeces will result in higher returns from the wool.

Skirting

The ideal procedure is as follows: Spread the fleece skin side down on a slatted or wire-topped table . Remove all manure tags and stained pieces and pack them separately. Never roll damp tags inside the fleece because they cause discoloration of any wool with which they come in contact. Separate the face and leg pieces from the fleece. Much more emphasis is required on the removal of these parts of the fleece when sheep have not been crutched. In the black-faced breeds, the face and leg areas usually contain black or grey fibres that are particularly objectionable to the manufacturer because they cannot be used in white or pastel-coloured goods. Burry, chaffy or straw portions should also be removed and packed separately.

Preparing the wool for market. After the fleece has been spread skin side down on a slatted or wire-topped table and the low-grade wool removed, one side of the fleece is folded in one-third of the way, then the other side is folded in to cover the first fold. The fleece is then rolled tightly from breech to shoulder.

Folding and rolling

When the low-grade wool has been removed, the most valuable portion is now ready to be folded and rolled. Fold in one side of the fleece one-third of the way and then fold in the other side to cover the first fold. Roll the fleece tightly from breech to shoulder to expose the best portion for inspection when graded . Packaging Black or brown fleeces should be kept separate, as should the tags and skirtings from such fleeces. When the fleeces have been folded and rolled, they are ready for packing in large jute wool bags to permit the wool to breathe. A handful of wool tied in each bottom corner will facilitate handling of the bags when they are filled. Mount each bag on a sacking stand, with the upper end supported by a ring that holds it open. Place the fleeces in the bag and tramp them in firmly. Tight packing permits maximum loading of shipping containers and facilitates handling. When the bag is full, release it from the ring and sew it with bag needle and cotton twine. One bag will hold approximately 30 fleeces and when filled will weigh between 110-160 kg. Storing the packed wool is an important consideration if it is not shipped to market immediately. Although wool can be held in storage for relatively long periods of time (if kept dry and protected from insects), it tends to deteriorate or lose its life after about two years of storage. For filling, the wool bag should be suspended on a sacking frame and the fleeces tramped in as tightly as possible. This permits maximum loading of shipping containers. Note the "ears" at the corners of each bag to facilitate handling.

Wool branding of sheep

Where branding is necessary, the sheep should be moved to holding pens as soon as they are shorn and marked with the owner's brand for identification. It is essential that the sheep be branded with a material that will not only keep the brand clearly legible for at least one year, but will also scour out in the processing of the wool by the manufacturers. Considerable damage to both machinery and materials can result from the use of an insoluble paint. Such damage increases the cost of manufacture and reduces the price paid by the manufacturer for wool. Only soluble branding fluids, which are available at all wool growers' supply houses, should be used for branding. A minimum number of brands should be placed on the sheep and the fluid used as sparingly as possible. Materials such as tar, lead paint and crankcase oil should never be used.

INDIAN WOOL INDUSTRY SCENARIO

In line with the trend in world production of different fibres, in India also, of the various fibres produced, raw wool accounts for a marginal share of 0.7% in total fibre production. India is the seventh largest producer of wool and contributes 1.8% to total world production. India ranks among the leading five countries in the world in sheep population, with a population of over 60 million sheep. However, while the world average for wool productivity has been about 2.4 kg/sheep/year, in India the average is 0.8 kg/sheep/year.

There are around 958 woolen units in the country, majority of which are in the small scale sector. The industry provides employment in the organised wool sector to about 12 lakh persons, with an additional 12 lakh persons associated in the sheep rearing and farming sector. Further, there are 3.2 lakh weavers in the carpet sector. In all, the total employment is about 27 lakh people.

The main wool producing states of India are Rajasthan, Punjab, Jammu & Kashmir, Karnataka, Gujarat, Uttar Pradesh, Uttaranchal, Andhra Pradesh, Maharashtra and Haryana. Punjab alone accounts for 40% of the woolen units, while Haryana accounts for 27%, Rajasthan 10% and the rest of the states account for the remaining 32%.

Types of Wool in India

The proper and systematic classification of wool with regard toils quality is not yet properly known but it is classified according to the territorial nomenclature as given here under:

Wool type	Colour
Joria Hanna	Superior
Bikaner	White White
RaJputana	Grey Super
Blbrik	White
Marwar	Yellow grey
Vicanere	White Grey
	Yellow grey
	Skin wool
	Common Balck

The management of wool production in India has been unplanned since long with the result that the persons who are engaged in wool production are not getting proper price. The Indian Council of Agricultural Research, New Delhi, has taken over the management of wool production and trading since 1937 and has appointed an authority, 'Sheep and Wool Development Officer' to control the whole production as well as the sale of the wool. The annual budget regarding the wool production is prepared by the officer and sent to the Govt. of India. The annual production of wool by different states has been recorded to be maximum from Rajasthan. The other states come on the position (Production-wise) as: Rajasthan > Uttar Pradesh > Punjab > Gujrat > Andhra Pradesh > Tamilnadu > Maharashtra > Karnatak > Jammu & Kashmir > Himachal Pradesh > Bihar.

The Government of India, for the holistic growth and development of Wool Sector, is making serious efforts to :

- Improve quality and quantity of wool (carpet grade, specialty wool fibres such as Angora & Pashmina, apparel grade and deccani grade wool). Improve fineness of wool by 2 micron and increase wool yield per animal by 10%.
- Reduce mortality rate of sheep and Pashmina goats by 10% to bring down mortality rate from current rate of 12-15% to 5-7 %.
- Setting up Common Facility Centres (CFCs) for creating processing facilities for wool and woolens.
- Increase employment in wool and allied industry and fulfill requirement of skilled manpower and also trained to wool growers with new techniques.
- Provide quality stud rams, pashmina bucks and angora rabbits as foundation stock to improve breed of good quality wool producing animals.
- Widen the uses of the coarse and colored Deccani wool (Southern region) by product development and product diversification.
- Improve wool production in rain fed areas by providing feed supplement to eligible (weak & pregnant) sheep and pashmina goats.
- Induce generic promotion of quality speciality fibres like Pashmina goats and angora rabbits.
- Benefit wool growers under Social Security Scheme of Govt. of India.
- Strengthen State Wool Marketing Organizations for marketing facility for raw wool and to ensure remunerative returns to wool growers.
- Increase the demand for Indian wool in domestic and international market.

PIG – PIGGERY

Classification

Kingdom: Animalia; Phylum: Chordata; Class: Mammalia; Order: Artiodactyla; Family: Suidae; Genus:*Sus*.

Introduction

The economic success of the industries, based on the animals and their products, depends on the proper production and usual development of the next generation of farm animals. A control over the future generations by scientific understanding of genetics, fertility, growth and development is the basis of the small scale-industries of farm animals. Besides goat, sheep, buffalo and cow farming another industry which is gaining success now-a-days, is called as 'piggery'. Pig has been used as diet of many tribes of Andaman and Nicobar. In India generally piggery is practised mostly by a few persons of lower classes of society. Now-a-days piggery is gaining good position among small scale industries in India because the pig is the cheapest and the best among the farm animals. The general appearance, habit and habitat of the pigs are well known, it would, therefore, be worthwhile to deal only the problems related with the high production of pigs.

Fertility and Fertilization

The oestrous cycle of pigs is completed within 20 to 21 days under normal conditions and the actual heat period lasts from 40 to 65 hours. In full grown pigs, vulva starts swelling from nine days before the actual heat day, after which it decreases in size and attains normal condition after nine days. This swelling is due to the oestrin produced by the ripening of follicle. In 40 to 50 hours of heat period, ovulation occurs after 18 to 36 hours from the beginning of the heat. During the heat period, if female pigs mate twice, the number of young ones produced are more than when they mate only once. Mating done at the end of heat period is not fertile. The low fertility of pigs is due to the foetal atrophy which is a genetic character of recessive nature and also causes reduction in the litter size in breeding pigs. The mature pigs during the lactation usually do not come under heat condition. It is found that if sperms produced by the boar at the time of mating are less, the size of the litter would be smaller. So, the boars should not be allowed to mate before three days of intervals. The mating period takes from 3 to 25 minutes. It is suggested for successful fertility that mating should not be disturbed for the complete ejaculation of sperms.

Artificial Fertilization

For the artificial insemination, well matured and healthy boar is allowed, to mate with a dummy female and semens are collected in artificial vagina with 2 pulsating pressure. The dummy female is better for the collection of semen in comparison to a living female. In such type of mating 200 cc. semen is produced by one male at a time. The semen is stored at about 40 to 50°F up to 4 days but fertility decreases after 24 hours of collection. The collected semen is diluted with 6 per cent solution of glucose before use for artificial fertilization. The semens saturated with carbon dioxide is sealed in ampules at temperature of 60°F, which gives 50 per cent result after a week of storage.

The litters produced by artificial insemination have the same number of young ones as in case of natural mating. In farm conditions, the litters produced by artificial insemination are about 50 per cent less than that of the normal.

Growth

In the pregnant pigs the anterior pituitary secretes lactogenic hormone during pregnancy which changes the metabolism of the mammary glands which start to produce milk. By the act of suckling, oxytocin is secreted and set free into the blood stream resulting into the formation and release of milk by the breast. The milk of pig is more concentrated than that of cow containing 10 per cent fat. The teats situated towards anterior part of the female yield more milk as compared to those at the hind end. The weight of the new born individual varies from 1 to 6 pounds. It is found that new borns are smaller in large litters than that of small litter. There is great individual variation in the production of milk due to number of young suckled, age and genetics of the sow. The maximum production of milk has been recorded in the third week of lactation. For the good growth of young ones, supplementary feed should be given after 25 days by creep feeding. It should be kept in mind that sows fed on synthetic diet may not be allowed to become much fatty otherwise the size of the litter would be smaller. The good growth in the early developmental stages plays very significant role for the development in length, muscle, weight of the pigs, so up to the age of 25 days young pigs should be fed on highly nutritive and easily digestible diet. The young pigs, with the weight of 50 pounds feeding on less nutritive diet, have good ability of fast growth to attain the maximum weight provided a good nutritive diet is given to them. The proportional growth of the body of pigs varies with the age i.e., the head and the legs are larger in comparison to the body but with the age, the body grows very fast as a result the muscles of thigh become concave. Such growth is found more in middle white pig in comparison to large white pig. It is very useful to have an idea about the change in the weight of the developing pigs and a proper control on that change.

The growth in pigs runs from the cranium towards backwards and from the tail forwards which meet in the lumber vertebrae as a result the loin region grows very fast followed by the pelvis and the thorax than the neck. The head grows very slowly.

It is advisable that the pigs in early stages should be allowed to feed on poor diet due to which early developing parts such as head and legs would grow very slowly and if later given, highly nutritive diet the parts which grow later i.e., loin would grow very fast and the carcass with high proportion of fat is produced. In the young pigs the fat is soft and in the older ones it becomes firmer when growth rate is very fast and far is synthesized from the carbohydrates.The pigs should be kept on starvation for 24 hours and given complete rest prior to killing to get standard product. The slaughtering of pig is carried out by making it unconscious by a hard stroke given on the head, and piercing a double edged kinfe in the neck to shed the blood out of its jugular veins. Now carcass is washed, cleaned well in hot water and cut open to separate the various parts for different preparations.

The products of Piggery are:

i) Pork

The flesh of pig is known as pork is general and the flesh obtained from different parts of the body have been given different names e.g., Bacon obtained from the back and sides and Ham from the back of the thigh.

According to the data produced by the Directorate of Marketing and Inspection, Government of India the production of pork is 5 per cent of total production of meat in India.

In addition to few Government factories a number of private concerns established in Delhi and Calcutta, produce and market pork and pork-products. The important pork producing factories are:

(1) Messers Essex Farms, Delhi, (2) Central Dairy Farm, Aligarh, (3) Elmac, Calcutta, (4) Foster, Bells, Gitwaco Farms and International Food Packers, Borivli, Bombay (Maharashtra).

ii) Bristles

The wiry and stiff hairs of the pigs, hogs and wild boars are known as bristles and they are obtained from the back and neck. Short bristles are obtained from the flanks and belly of the animal. The bristles obtained from the living pigs are superior in quality than those of slaughtered ones. The rough and coarse bristles are generally used for varnish work and painting brushes.

The main bristle producing areas are Uttar Pradesh, Madhya Pradesh and Punjab. The principal trading centres are Kanpur and Jabalpur but 70 per cent of the total bristle export is being made from Kanpur.

Two types of bristles viz., Desi bristles obtained from pigs of Uttar Pradesh and Madhya Pradesh, and. Darjeeling bristles, obtained from Himalayan foot hill and Darjeeling district. The bristles of three colours are usually found viz., white, grey and black.

The important bristle markets in the country are at Agra, Allahabad, Azamgarh, Faizabad, Jaunpar; Amaraoti, Nagpur, Santhal Parganas, Calcutta and Kakinada.

Bombay is the biggest port to export the bristles worth millions of dollars to U.S.A., U.K., West Germany and Japan. Indian Standard Institute, New Delhi standardizes the bristles and provides AG MARK. Only the properly graded bristles under 'Bristle Grading and Marketing (Amendment) Rules-1962' are permitted for export purposes.

iii) Sausages

Sausages are prepared by fresh minced pork, free from bone and skin. For the preparation of sausages usually shoulder piece, ham and bacon are preferred. The sausages are prepared by washing, cooking and mixing the pork with spices such as white pepper and paprika to make it delicious.

iv) Lard

Lard is the fat of pig, squeezed from the body tissue is termed as lard. For the preparation of lard, carcasses are cut into pieces, minced, boiled over a furnace and the fat is removed carefully when it is boiling hot.

Lard is used as a fine cooking medium and in the manufacture of soaps, lubricants, greases, candles and water proof materials.

v) Other Uses

The most important pharmaceuticals derived from thyroid, pituitary and pancreas of pigs are pepsin, thyroxin, pituitrin, insulin, liver extract, testosterone etc. The pig toe-nails are used for making tobacco fertilizer, plaster and other plastic materials. The blood is used as food for livestock and poultry and also as manure.

COW – DAIRY

Classification

Kingdom: Animalia; Phylum: Chordata; Class: Mammalia; Order: Artiodactyla; Family:Bovidae; Genus:*Bos*; Species:*indicus*

Introduction

Man uses mammalian milk for a variety of preparations like curd, butter, cheese, sweet etc. and hence domesticated a number of mammals like cows, goats and buffalos. The other animals like sheep, camels, asses and mares are milked in certain confined localities but as producers of milk they are of little importance.

The dairy cattle thrive best in areas where pasturage and other green forage are grown in abundance. Extremely cold climates are not suitable because of the lack of green forage and much expense for protecting the animals from the weather. The dairy industry involves production, processing and distribution of milk and milk products. The adoption of pasteurization and the enforcement of laws requiring proper food value in dairy products greatly benefited the entire dairy industry.

The number of well recognised cow breeds in India is about fifty. In addition, a large number of other types which do not confirm to any defenite breed characteristics exist, and are treated as non descripts. The non descripts are very poor producer of milk.

The principal breeds of cows are Haryana,.Kankrej, Ongole, Rath, Deoni, Gir and Kangayam. Deoni is found in the locality of North-western and western parts of Hyderabad and are good milkers in the region. Gaolao is common in Wardha and Chindwara districts. Gir is good milker and found in Kathiawar, Rajputana and Baroda regions. Haryana is very good milker found in the vicinity of Rohtak, Hissar, Karnal, Delhi and Uttar Pradesh. Kankrej is common in South east of Rann of Kutch and Ahmedabad region and is fair milker. Ongole is also a good milker and found in Guntur districts. Rath is fairly good milker which is found from Rajasthan to North-western part. Shahiwal and Sindhi are very good milker found in Punjab, Haryana and Uttar Pradesh.

Breeding

Near about the middle of the 18th century, dairy farmers began the improvement of cattle and other farm livestock The chief practices followed were the mating of related animals and close culling. The rearing and maintenance of proved and standard breed for distribution upto village level is undertaken in Government farms. The bulls are used for selective breeding in areas of well-defined breeds. Attempts are being made to upgrade non-descripts in several areas by repeated forward crossing. These attempts have given encouraging results but the number of bulls available for breeding purposes are inadequate. The progeny of Shahiwal, Sindhi, Haryana or other well defined bulls, yield atleast twice as much milk as the non-descript dam. The second cross shows a further increase in the production of milk. Considerable amount of cross-breeding using imported bulls, has been carried out during the last 50 years. In the earlier years, the breeds favoured in India were the Ayreshier. A large number of Friesians were later imported which are reported to have given best results. The Indian Friesian yields on an average 4000 Kg of milk per lactation as compared to 5000 Kg given by Friesian cows. The Friesians are regular calvers, their productivity, however, is dependent on the care and attention given to them. These are maintained in favourable areas and moved to the hips in summer. The yield goes down as the percentage of Friesian blood is increased or decreased. It has been found that an increase in the production of Friesian blood increases the capacity to produce but the constitution of animal is not equal to the strain imposed in producing the milk to capacity. Although cross-breeding with foreign breeds improves the milk production, it is not widely adopted as the animals need expert care and management, not usually given to cattle in India. Experience shows that the cow produced by repeatedly back-crossing the half-bred is superior to the cow started with. For more effective utilization of limited number of breeding bulls now available, artificial insemination has been introduced throughout the country. In the very beginning the artificial insemination was followed at regional research centers in Calcutta, Patna, Montgomery, Bangalore and Izatnagar.

Feeding Stuffs

The feeding stuff for dairy cattle may be broadly 'classified into roughages and concentrates. The roughages consist of succulent feeds (natural grazing, cultivated grasses, cultivated fodders and root crops) and dry fodders (hay, straw, chaff). The concentrates consist of carbohydrates-rich materials (oil seed, oil seed cake and meals). In addition to roughages and concentrates, dairy animals also require a certain amount of common salt to keep them in good condition.

Investigations in this area show that a number of materials hitherto considered as wastes, can be utilized as cattle feed eg. mango seed, jaman seed, and mahua flowers (concentrates) and ground nut husk, bajra and coffee. husk (roughages). In addition to distillary products (malt sprout, millets and molasses) feeds of animal origin eg. fish meal, blood meal and bone meal may be employed as cattle feed. The table below gives information about the feeding stuff, commonly fed to dairy cattle in India.

Dry roughages	Green fodder	Concentrates
Wheat straw	Maize	Gram
Oat hay	Lucern	Bran
Rice straw	Berseem	Cotton seed
Maize stalks	Oat	Cotton seed cake
Juwar stalks	Juwar	Mustard cake
Legume straw	Bajra	Linseed cake
Grassi hay	Elephant grass	Til cake
Bahra stalks	Ground nut cake	
Berseen hay		

Maintenance ratio

A well fed dairy cow producing one pound of butter fat per day utilizes about 48 per cent of the food consumed for maintenance. If the ration is cut down to about two thirds, the cow will still need the same quantity of food for maintenance, but the quantity available for milk production is reduced.

The results obtained from maintenance rations trials on cow show that animals fed on green fodder and hays remain fit for a long time with a positive nitrogen balance.

The maintenance requirements of dry cows (wt. 400-500 Kg) and heifers (400 Kg) show that they should be given fodders of oats -20 to 25 Kg, barseem -35 to 40 Kg, maize - 30 to 35 Kg, bazra -25 to 30 Kg, elephant grass -30 to 40 Kg and sun flower - 25 to 3O Kg, per day. The dry roughages like, oat hay - 5 to 6 Kg, berseem hay - 6 to 7 Kg and maize - 7 to 8 Kg should be given daily. The data given regarding the composition of rations for the maintenance of dairy cow are of two category. The ration first is constituted of green maize - 30 Kg, wheat bran - 2 Kg, gram - 1 Kg and toria cake -1/2 Kg in which the total digestible nutrients is 7 to 8 Kg and digestible protein is 1/2 to 1 Kg. The ration of second category is constituted of barseem - 30 Kg. wheat straw - 3 Kg, bran - 1 Kg and maize -3 Kg in which the total of digestible nutrients is 7 to 8 Kg and the digestible protein is 1 Kg.

Diseases

The main diseases of dairy cattle in India are rinderpest, haemorrhagic septicemia, black fever, anthrax, food and mouth disease, three-thy fever, pleu-ropneumonia, tuberculosis, jone's disease, tie fever Surra, tetanus, rabies, calf diptheria etc. The vaccines and antisera for cattle diseases are produced at various centres in the country. Many of the above mentioned diseases are under control, if proper treatment is given at proper time. Veterinary doctors have been appointed to look after the cattle in villages but due to lack of proper knowledge and facilities dairy animals in the villages are not being cared properly in general routine and illness.

Milk

Milk is produced by the mammary glands, which are specialised skin glands. The actual secretion of milk by the mother is stimulated at birth by a lactogenic hormone (galactogen or prolactin) which comes from pituitary gland located at the 'bottom of brain, adrenal hormones also are essential for lactation. During gestation, the production of lactogenic hormone from pituitary is inhibited by the presence of another hormone, estrogen, which disappears at birth. In some mammals the stimulus of sucking offspring serves, through the nervous system, to stimulate the secretion of lactogen.

Milk is essentially an emulsion of fat in a continuous phase. The disperse phase consists of fat globules of varying diameters from 0.1 to 10%. The number of globules varies from 2.5×10^9 to 5×10^9 per centimetre. The globules are surrounded by layers of protein, phosphalipids and probably also carotene and cholesterol which prevent them from coalescing together to large globules. The protective film may be broken by churning the milk when the fat globules coalesce to form butter. Milk fat or butter consists of glycerides mainly of butyric, caproic and capric acids which possess characteristic odours. The continuous aqueous phase consists of carbohydrates (almost entirely lactose) and a colloidal suspension of casein stablized by lactalbumin and lactoglobulin. Lactose forms the largest single constituent of cow's milk, next only to water. Cow's milk contains 2.4 to 6.1 per cent lactose.

Milk also contains native enzymes (those present in udder) and enzymes originating from bacterial contamination. Amylase, catalase peroxidase, lipase, phosphatase, galactase, lactase and aldehydase are among the enzymes present in the milk. The activity of most enzymes present in milk is destroyed or greatly reduced by pasteurization. The white appearance of milk in reflected light and its opacity in transmitted light are due to the emulsified fat and the colloidal calcium phosphate and caseinate. The creamy colour is due to the presence of carotene in the disperse phase and of riboflavin in aqueous phase. The flavour of natural milk is pleasant and slightly sweet. Milk containing 3 to 5 per cent fat and rich in lactose percentage possesses a better flavour.

As regard to the nutritive value, milk is a complete food for infants up to 6 months of age, after which it acts as a supplement to other food. Cow milk is easily digestible. Fifty per cent of its caloric value is contributed by fat, 20 per cent by lactose and 21 per cent by protein. Milk is a good source of phosphorus, calcium and vitamins. Milk is rich in vitamin A, excellent source of vitamin B2 and fair source of vitamin B1. It is deficient in vitamin C.

Due to its high nutritive value and fluid nature, milk is an ideal medium for growth of bacteria and it is of prime importance that handling and processing of milk should be carried out under strictly hygienic conditions. Milk contains bacterial flora normally associated with the udder and those introduced during milking. All contaminations should be effectively removed.

Processing of milk

In advanced countries a major part of the milk is processed and rendered safe before distributing to consumers. Among the methods employed, pasteurization is the most important. This consists of applying heat for the minimum period necessary to destroy microbial contaminations, the temperature being high enough to kill them but not so high as to affect the chemical and physical characteristics or the nutritive value of milk.

Two methods of pasteurization are in use i.e. the holding and flash High Temperature Short Time (H.T.S.T.). In the holding method, the milk is heated from 140 to 150°F, held at that temperature for 30 minutes and rapidly cooled to 50°F. In the HTST method, the milk is heated from 160 to 162°F, held at that temperature for 10 to 20 seconds and cooled immediately to 50°F. Holding pasteurization, gives a safe and satisfactory product. Pasteurization does not affect the concentrations of vitamin A and D or riboflavin. About 25 percent of thiamine and 50 percent of ascorbic acid are destroyed. Among the other milk processing methods the irradiation by ultra-violet light, developed in Germany during world war IInd, is of interest. Irradiation increases the vitamin D content of the milk.

Homogenizing and freezing processes were developed in America. The homogenizer is a pump which is designed to subject a liquid milk product to a higher pressure, then allows the product to pass through a special value. This process breaks up the fat globules, increases the ability of the protein to hold water of hydration and slightly softens the curd. After homogenization the milk passes on to the heating section of the pasteurizer. The bulk of milk is marketed in India in the raw condition. Pasteurization has not been widely adopted in India owing to the difficulty of collecting milk at a reasonable price from rural areas and of finding a market for pasteurized milk.

The pasteurization plants have been set up in organised dairies throughout the country. The holding method is generally, employed for despatch to camp or hill depot having a temperature upto 50°F for local sale. Where refrigeration plants are not easily available, pasteurized milk is cooled to 60-65°F. in cold weather by ordinary water. Some of the dairy forms sell pasteurized milk in sterilized bottles. The bottles are washed in washing machines and passed to filling machines. They also make caps from covering strips and seal them on to the bottles. The filled bottles are kept under cold storage. They are taken out for distribution.

Marketing and distribution.

In urban areas 60 to 70 percent of the total milk requirements is produced within the municipal limits, the rest is obtained from adjoining rural areas. Only 6 to 8 per cent of the total milk produced in the country is transported from rural to urban centres for consumption as milk and milk products. Nearly 2/3rd of milk teceived from outside the municipal limits comes from within 8 to 15 kilometres of the towns, and the remaining 1/3rd from beyond this distance. A part of the milk consumed in large cities like Calcutta, Bombay, Madras, Delhi and others, it obtained from localities situated at a distance of even 75 kilometres. Some successful efforts have been made to organise the production and marketing of milk on a cooperative basis.

Prices

The price of milk varies to a great extent from place to place. In rural areas there is practically no good market facilities for fluid milk. The milk left over after meeting the demands of the producers family is converted into butter, ghee or khoa. These products are sold at weekly markets at prices largely determined by the distance of the market from the village. As a rule, cow milk is cheaper than buffalo milk but it may not be true for all the time and for all the places.

Milk Products

A variety of milk products is known in India and some of them figure in inter-state trade. Dairy products such as cheese, butter, condensed milk, milk powder, curd etc. make the dairy farming a highly attractive industry.

Dahi (Curd)

It is prepared by souring milk with a lactic acid starter which is usually curd of the previous day. For the preparation of curd, milk is first boiled and cooled from 60 to 70°F. Further, starter is added and is kept untouched. Thus, curd becomes ready for consumption in 10 to 12 hours. Curd contains 0.6 to 1.0 per cent titratable acid expressed as lactic acid. It has a lactic flavour and a compact smooth texture. Its composition varies with the quality of milk used, the types of organisms present in the starter and the time allowed for souring. The organisms present in the curd are mainly *streptococci* type and *Lactobacillus casei*. Curd is normally consumed at the place of production. Curd contains 84.79% water, 7.7% fat, 3.4% protein, 4.6% lactose, 0.7 to 0.8% lactic acid, 0.7% minerals, 0.12% calcium and 0.95% phosphorus.

Cream

Cream is a fat containing fraction separated from milk by centrifuging the liquid milk. The separator, commonly employed, consists of bowl with a large number of conical discs arranged one above the other with intervenning spaces. Milk enters through an opening in the centre and as the bowl is rotated, 3,000 to 20,000 rpm, the lighter fraction which is cream, is driven towards the centre and the heavier fraction or skimmed milk is drawn towards the periphery and drawn out.

The second method to obtain the cream is gravitational method in which milk is kept in a container at 50°F. After 24 hours cream automatically comes on the surface of the milk Which is taken out by spoon. The yield and quality of cream vary according to the quality of milk and to the speed of separator. Skimmed milk is obtained as a by product which contains 0.04 to 0.5% fat and is used in the manufacture of condensed milk, milk powder, butter milk and cheese. The yield of cream is about 10% from buffalo milk, 6% from cow milk and 7.5% from mixed milk. The composition of cream is 56% fat, 1.6% protein, 3% lactose, 0.4% minerals and 39% water.

Butter

Butter is a mixture of milk fat, butter milk and water. Salt and colouring materials are often added. It is good source of vitamin A and fair source of vitamin D. Butter is characterized by spreadability, a characteristic not found in butter substitutes. This probably is due to the glyceride structure of butter fat and to the presence of lower saturated fatty acids.

Desi butter or makhkhan is prepared by churning curd after dilution with water, in earthen or tinned metal pots by a wooden pole to one end of which beaters are attached which is known as "mathani". This butter usually contains 18 to 25% water and varying quantities of curd.

The composition of butter is 14% water, 83.5% fat, 1.5% lactose, 0.3% minerals and 0.8% aluminium.

Creamy butter

The production of creamy butter in India is confined to few dairies. For the Preparation of creamy butter, cream is aerated and ripened at 75 to 95°F to develop the desired smell and to facilitate churning. The ripening is effected usually by the addition of natural starter such as curd, fermented cream, and butter milk or commercial starters containing pure cultures of lactic acid bacteria. The ripened cream, after dilution with cold water is mixed with a small quantity of dye and then churning is started at temperature 50 to 60° F. Churning is usually carried out in the mornings, when temperature is low otherwise ice is added to maintain the desired temperature. The whole operation takes 30 to 40 minutes. Butter milk is drained off and the butter is washed repeatedly with cold water. It is then salted, up to 2.5%, by adding salt solution either in the churn or after taking out the butter. Excess of moisture is pressed out to give a product with a firm and compact texture.

Ghee

Ghee or clarified butter is obtained from butter by eliminating water. It is next to milk in importance as a dairy product. It can be stored over a long periods and for this reason it is preferred in comparison to butter in most of the tropical countries for common people. Ghee is produced in India according to the traditional process involving sour curling of milk, recovering butter, and heating the butter to remove water. Desi butter is preferred to creamy butter as ghee obtained from the former is melted into ghee at once or after storage upto 10 days. It may be partially dehydrated and later converted into ghee according to the demand of market. The temperature employed for clarifying butter varies from 80 to 125°C. Short exposure to 120°C does not interfere with the formation of grain nor does it diminish the carotene and Vitamin A contents of the ghee. To ensure proper grain formation, the tins to which ghee is transferred should be kept undisturbed and cooling should be allowed to take Place gradually. The appearance of colour and grain structure influences its market value. The ghee of cow is yellow and that of buffalo is whitish. The grain is buffalo ghee are bigger than that of cow ghee. The composition of the milk from which it is derived. Ghee is mainly used for cooking, frying and taking directly with ithe food materials. Ghee is subjected to extensive adulteration in the trade.

Malai

When milk is heated, a layer of fat and coagulated proteins, malai, is formed on the surface, slow heating helps to increase the thickness of the layer: The volume of malai can be increased by boiling the milk until a voluminous froth is formed and cooling slowly over a dying fire. Malai is either consumed directly, or used in the preparation of sweets. Its composition is moisture 60 to 70%, fat 25 to 30%, proteins 3 to 3.5%, lactose 3.3 to 3.8% and ash 04 to 0.5%.

Condensed milk

It is obtained by evaporating milk at 130 to 135°C in a vacuum pan to the required concentration. The concentration is homogenized to prevent the separation of fat, cooled and fortified if necessary. Stabilizers such as disodium hydrogen phosphate or calcium chloride, are added to prevent coagulation during the sterilization. The condensed product is cooled rapidly from 80-86°F and held at that temperature for 15 to 20 minutes. The cooling is so controlled that the crystals are of small dimensions and remain in suspension in the viscous liquid.

Khoa

It is prepared by the rapid evaporation of water from the milk. It is usually prepared from buffalo milk by heating with brisk stirring in flatbottomed shallow steel pots until the volume is reduced to about one-fifth. The product is gathered in a compact mass, cooled, and packed for markets. Alum is sometimes added to the milk during the boiling to give a smooth texture to the product. Khoa is consumed directly or used as an ingredient of sweets. It can be kept for 3 to 4 days without deterioration. The common adulterants of khoa are cereal flours.

Cheese

Cheese is the product made from the curd, obtained from whole or skimmed milk with or without added cream by coagulating the casein, and then further the separated curd is treated by ripening ferments.

Soft cheese, known as Paneer' is prepared by using coagulants as the source of coagulating enzyme for clotting milk. Milk is warmed to about 100°F and the crushed coagulants are tied in cloth and dipped in it. The milk curdles in 30 to 40 minutes. The coagulum is placed on a muslin cloth and the whey is drained out. The process of cheese manufacture vary very much but not to such an extent as may be with different characters of the final product.

Future prospects

The present retail prices of milk in urban areas in India are higher than those in the other countries. This is due, principally, to inadequate supply of milk. With a view to increase the per capita consumption of milk, the Planning Commission has suggested the adoption of measures for raising milk production in suburban areas. Emphasis has been laid on the need for maintaining hygienic conditions during the collection, transportation and distribution of milk and for enforcing measures of quality control. The setting up of a statutory milk board for each urban area, consisting of representatives of producers, distributors, consumers, municipal corporation, health authorities and the state Governments is a good initiative from Planning Commission. The matters related to the handling, distribution, quality control, imports and price of milk, and milk products come under the juridiction of the board. The financial assistance needed will be provided by Government, municipalities and cooperative banks. Village schemes have been formulated for improving the breeds of milk cattle and for ensuring adequate fodder supplied for dairy cattle.

AQUAPONICS – AN INTEGRATED FARMING WITH SYMBIOTIC RELATIONSHIP

Introduction

Aquaponics is a typical kind of biologically integrated system as it forms a vital link amid the continuously recirculating aquaculture with hydroponic floral production. A type of symbiotic relationship is established between aquatic fauna with that of plants within the system. Current advances by researchers all around the world have arched aquaponics into a sustainable working model for food production. This innovation termed aquaponics respects the principle of sustainability and also gives a prospect to increase the economic efficiency with an added productivity. It's high time that we need to reconsider the agricultural sciences in a way that can develop technologies in an eco-friendly manner.

Aquaponics is the fusion of aquaculture or fish cultivation and hydroponics or plant farming devoid of soil. The increasing rate of scientific and technological innovation has kept researchers in a continuous struggle to update themselves with the most up-to-date codes of practices, technologies and scientific breakthroughs. The need and necessity of sustainable development for the aquaculture is beyond the thought. Improved productivity with condensed ecological impact, integration between production systems and reduced use of chemicals are some of the leading principles that more sustainable fish production needs to follow (Diver S, 2006). The safety of food for human utilization is frightening on a worldwide level. Aquaculture represents fish farming, system where commercial fishes are reared in containers, ponds or tanks.

Hydroponics refers to the production of plants without soil. Plant roots grow in a nutrient solution with or without an artificial medium for mechanical support (Pantanella E, 2008)

Hydroponics is one of the several plant culture techniques, which enables plant growth in a nutrient media with the mechanical support of inert substrata. Hydroponics is considered as one of the promising technique not only for plant physiology experiments but also for commercial production (Hutchinson W, 2004).

Both aquaculture and hydroponics have some pessimistic aspects. Hydroponics requires expensive nutrients to feed the plants, and also cyclic flushing of the systems is required which leads to waste clearance aspects. Aquaculture demands excess nutrients to be removed from the system; usually this means that some amount of the water is removed, generally on a day by day basis. This nutrient rich water needs to be disposed off regularly and replaced with clean fresh water. While aquaculture and hydroponics are both very competent methods of producing fish and vegetables, but when we look at combining these two, these negative aspects are curved into positives.

Fish generates mainly nitrogenous wastes. If these wastes mount up, it can be lethal for the life of fish, but on the other hand, if they can be managed efficiently then the same waste can act as a great fertilizer for plants. As the plants consume these nutrients, they cleanse the water, which is favorable for the fish consumption. Many cultures have been made using this cycle to grow improved crops and rear the fish as a supplementary food source. This easy logic is the base for Aquaponics culture. The Aztecs developed a method of building floating islands for food-plants such as maize and squash. Fish use to move around the islands, leaving their waste on the lake bottom, where it could be collected to fertilize the plants.

Modern aquaponics is somewhat more technically competent which makes use of environment friendly approach to generate food. Fish are habitually kept in voluminous tanks and the plants are grown hydroponically; that is, without soil. Plants are generally planted in beds with a few gravel or clay and their roots hang down into the water. The water is recycled through this system, so that it collects the "waste" from the fish and recirculates back to the plant beds, where it is naturally filtered by the plants and then again returned to the fish tanks. In this fashion of culture, no chemical fertilizers are required for the plants unlike traditional farming techniques as in the present context they all come from the fish-waste. It also tends to be organic, since the use of pesticides would be harmful to the fish.

Thus, aquaponics is a sustainable system that combines both hydroponic (plant) and aquaculture (animal) systems. This system makes use of the natural biological cycles (Nitrification). It allows us to generate fish and plants in a single system with a large diminution in water use (Fig.1)

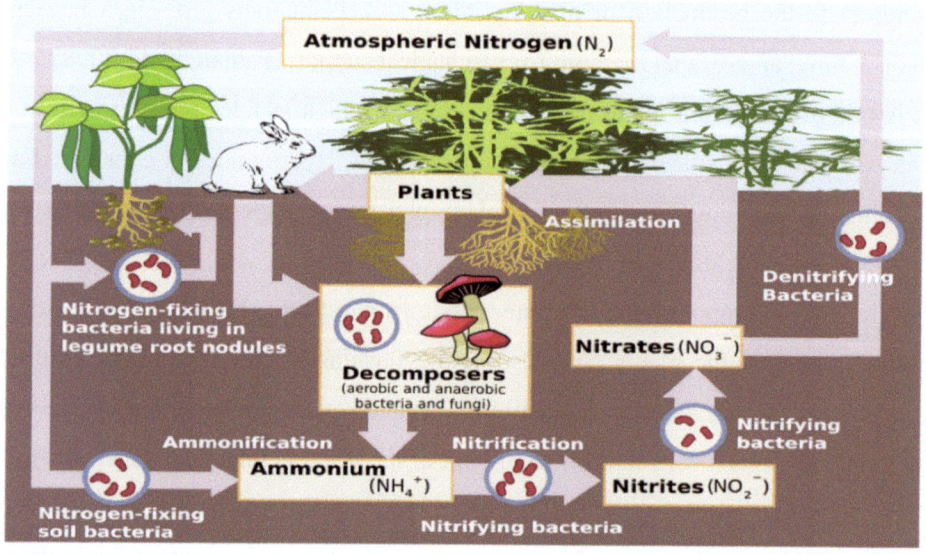

Fig.1: Aquaponics basic diagram

Why aquaponics?

Aquaponics sufficiently helps to negotiate many of the crises widespread across the globe. Some of the universal crises are increasing population, food shortages, increasing unemployment, global warming, etc. This system uses a fraction of the water, about 10% of soil growing. There is no necessity to purchase, store and apply fertilizer, no soil-borne diseases, no tilling, and no weeds. It results in high fish stocking density, high crop yield. This integrated system relies on the principle of no waste as waste from fish is been used by plants. Water is being re-used in the re-circulating system. No pesticides or herbicides required rather continuous organic fertilizer is supplied naturally. This system aids to food security as we can grow our own food within a defined space, year-round and equally potent in draught or places with poor soil quality which results in local food production, enhances the local economy and reduces food transportation. Thus, aquaponics can be considered as sustainable as it has lots of advantages with respect to hydroponics and aquaculture (Table 1) along with a cutting edge for meeting a number of crises.

System	Advantages	Disadvantages
Hydroponics	Produces a high volume of crops in a small space The most water efficient method of crop production	Dependent on manufactured fertilizers that are costly
Aquaculture	Produce a large volume of fish in a small space	It has a high rate of failure due to high stocking rates Fish produce ammonia, algae, minerals that are to be constantly filtered
Aquaponics	No pesticide, thereby reducing carbon footprint The plants get an automatic food supply from the fish water The plants filter the water for the fish	Management requires skills in growing fish and plants

Table 1: Comparison between hydroponics, aquaculture and aquaponics.

How Aquaponics Works (Fig. 2)

Fish are raised in a tank

Water from the fish tank is pumped to the plants

Bacteria convert ammonia and nitrite to nitrate

Plants absorb the nutrient rich water

Filtered water is returned to the fish tank

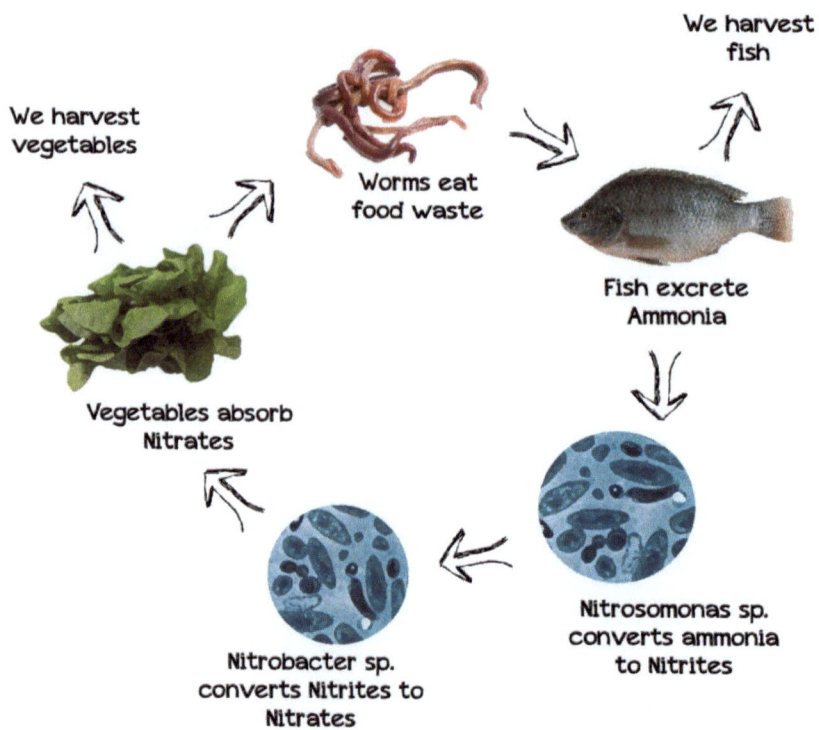

Fig. 2: Diagram illustrating general working of Aquaponics

Components

Aquaponics comprises two main parts, aquaculture part for raising aquatic animals and the hydroponics part for growing plants (Rakocy, Diver, 2006). Although consisting chiefly of these two parts, aquaponics systems are usually catagorized into several components or subsystems responsible for the effective removal of solid wastes, for adding bases to neutralize acids, or for maintaining water oxygenation (Rakocy, 2006). Typical components include:

- Fish Tank
- Place to Grow Plants
- Water Pump(s)
- Air Pump
- Irrigation Tubing
- Water Heater (Optional)
- Filtration (Optional)
- Grow light (Optional)
- Fish and Plants
- Sump
- Settling basin

Fish tanks are used to rear fish. Water pumps and air pumps are used to regulate the water level and air level respectively. Irrigation tubing is well connected throughout, so that it creates a re-circulating system. Sump is the lowest point in the system where the water flows to and from which it is pumped back to the fish tanks. Settling basin is a unit for catching uneaten food and detached biofilms, and for settling out fine particulates (Fig. 3)

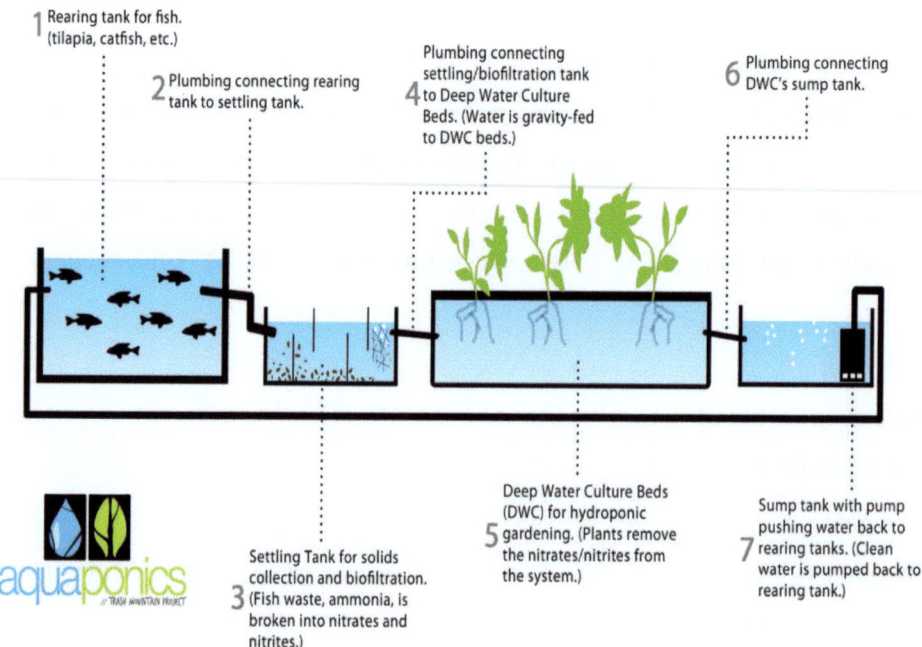

Fig. 3: General setup of an aquaponics system

Plants: hydroponics

Plants are grown as in hydroponics systems, with their roots immersed in the nutrient-rich effluent water. This enables them to filter out the ammonia that is very toxic to the aquatic animals, or its metabolites. Once the water has passed through the hydroponic subsystem, it is cleaned and oxygenated, and can return to the aquaculture vessels. This cycle is constant and continuous. Devoid of plants the system cannot function properly. Growing plants in soil is quite easy but takes up valuable space because of moisture and spacing requirements.

Aquaponics takes care of this involuntarily, without much contemplation except to insure the flow of water. If at all the electricity quits or a pump fails, the plants will survive several days up to two weeks depending on the temperature, but of course this condition may be detrimental for the survival of fish. Even plants needing large amounts of nitrogen, like tomatoes, can exist side by side with plants that require little, like lettuce. The nutrient rich water reaches all plants and because it only passes through, only what is needed is used. Even with good plant coverage there are a lot of nitrates flowing out the drains back to the fish tank, enough in fact to power up another group of grow beds. This is not a concern unless the water is cloudy in the fish tank.

Vegetables like Lettuce, Beans, Squash, Zucchini, Broccoli, Peppers, Cucumbers, Peas, Spinach,,etc. Herbs like Basil, Thyme, Cilantro, Sage, Lemongrass, Wheatgrass, Oregano, Parsley, etc. Fruits like Strawberries, Watermelon, Cantaloupe, Tomatoes, etc. Most garden varieties flowers can also be grown.

Why do Plants like Aquaponics?
- Nutrients constantly provided
- Warm water bathing the roots
- Don't have to search for water or food
- Less effort needed in putting out roots
- All the energy goes into growing UP not DOWN
- No weed competition

All of the above mentioned factors support all the necessary ingredients responsible for a improved and healthy growth of the plants.

What influences the amount of available nutrients to plants?
Several factors as mentioned below held accountable for the availability of nutrients to the plants. These factors should be tightly and timely regulated for the best possible growth of the plants.
- Density of fish population
- Size of fish
- Temperature of water
- Amount of uneaten fish feed in water
- Availability of beneficial bacteria
- Amount of plants in the system
- Media present in system
- Water flow rate

Economical rising effectiveness with vegetables production
Aquaponics presents a promising opportunity to reorganize the traditional fish farming, to obtain in more money at the farm gate. Two profit centers for producers: fish and plants. If fish goes through a small cycle then we have plant revenue to rely on and vice versa. The integration of fish and plants is a kind of polyculture that increases diversity and thereby enhances system strength. Aquaponics increase economical efficiency because several key costs such as nutrients, land and water are substantially reduced and module operating and infrastructural costs are mutual.

The system involves no control of root pathogens, as these are controlled biologically by the broad spectrum of antagonistic micro-organisms that develop in the natural environment

(Nichols M, 2008). Aquaponics is a bio-integrated system that associates recirculating aquaculture with hydroponic vegetable, flower, or herb production (Gordon A Chalmers, 2004). This production type of fish and vegetables is right where the market is headed-consumers are demanding safe food produced in an environmentally responsible way. Aquaponics process, gives big advantages in earlier and faster plant crop production to capture more profits. This type of agriculture might mean a stepped-up investment, but it is one that creates another revenue stream (from fish) linked with more profitable plant production. Some benefits of this system outlined by Amadis Lacheta (2010):

- Faster growth rate, crop maturity and yields
- Consistency and quality of crops
- Drastically reduced water and nutrients compared with soil-grown produce
- Crops can be grown in places where ordinary horticulture and aquaculture is impossible due to poor or contaminated soil or water
- Reduced growing area required
- Systems can be set up at a comfortable working height, excellent for people who are elderly or have disabilities
- Relative freedom from soil diseases and pests
- Weeds are virtually non-existent
- Water stress is reduced in hot conditions
- Less ongoing maintenance required

Increasing economical efficiency of aquaculture by aquaponics, is given from the fact that by this innovation water consumption is reduced to least amount and most important is, we obtain organic vegetable products, that means an additional product which brings to us additional cash.

Animals: aquaculture

Aquariums require filtering systems that must be either cleaned or replaced on a usual basis. The grow beds of the aquaponics system by themselves act as the filter without the hassle of cleaning or replacing. Of course, plants must be present in the grow beds. Almost a lot of freshwater fishes can be raised in the system although the operating temperature may prohibit rearing of some species such as trout. Freshwater fish are the most common aquatic animal raised using aquaponics, Fish like aquarium fish, Tilapia, Trout, Catfish, Yellow Perch, Bass, Bluegill, Carp, Koi, Goldfish, freshwater Prawns are recommended for rearing in an aquaponic system.

Fish Maintenance
- Feed fish 2 - 3 times a day, but shouldn't be overfed
- Fish eat 1.5 – 2% their body weight per day, this should be taken care of
- Fish should be fed only that which they can eat in 5-10 minutes
- Fish won't eat if they are too cold, too hot or stressed, thus temperature conditions should be well regulated
- Water quality should be checked periodically
- Fish behaviour and appearance should be observed

Fish Health Management
- Good hygiene and bio security—prevention, avoidance, selective access, and commonsense should for all time be exercised.
- Before stocking fish from other facilities into own's system it should be quarantined properly. Their health should be monitored for quite a few days—treat if necessary.
- The best defense is fish's own immune system. Always there should be a low-stress environment so that fish will maintain their health.

Bacteria

Nitrification which involves the aerobic conversion of ammonia into nitrates is one of the basic functions in an aquaponics system as it adds in reducing the toxicity of the water for fish, and thus allows the resulting nitrate compounds to be removed by the plants for nourishment (Rakocy, 2006). Ammonia is steadily released into the water through the excreta and gills of fish as a product of their metabolism, but must be filtered out of the water as higher concentration is unfavorable to fish. Although plants can absorb ammonia from the water to some degree, nitrates are assimilated more easily thereby efficiently reducing the toxicity of the water for fish (Rakocy, 2006). Ammonia can be converted into other nitrogenous compounds through:

- Nitrosomonas: bacteria that convert ammonia into nitrites, and
- Nitrobacter: bacteria that convert nitrites into nitrates.

In an aquaponics system, the bacteria responsible for this process form a biofilm on all solid surfaces throughout the system that are in constant contact with the water. The submerged roots of the vegetables have a large surface area, so that many bacteria can accumulate there. Care for these bacterial colonies is important as to regulate the full assimilation of ammonia and nitrite. This is why most aquaponics systems include a biofiltering unit, which helps facilitate growth of these microorganisms. As the nitrification process acidifies the water, non-sodium bases such as potassium hydroxide or calcium hydroxide can be added for neutralizing the water's pH. In addition, selected minerals or nutrients such as iron can be added in addition to the fish waste that serves as the main source of nutrients to plants (Rakocy, 2006).

A fine way to deal with solids buildup in aquaponics is the use of worms, which liquefy the solid organic matter so that it can be utilized by the plants and/or animals.

Technical Operation

Dr. James Rakocy, the director of the aquaponics research team at the University Of The Virgin Islands had issued ten key guiding principles for creating successful aquaponics systems based on extensive research done as part of the Agricultural Experiment Station aquaculture program:

- Use a feeding rate ratio for design calculations
- Keep feed input relatively constant
- Supplement with calcium, potassium and iron
- Ensure good aeration
- Remove solids
- Be careful with aggregates
- Oversize pipes
- Use biological pest control
- Ensure adequate biofiltration
- Control pH

The fundamental inputs to the system are water, oxygen, light, feed given to the aquatic animals, etc. In terms of output, an aquaponics system may frequently yield plants such as vegetables grown in hydroponics, and edible aquatic species raised in an aquaculture. Typical build ratios are .5 to 1 square foot of grow space for every 3.8 L of aquaculture water in the system. 3.8 L of water can support between 0.23 kg and 0.45 kg of fish stock depending on aeration and filtration. Target pH should be maintained between 7.0 – 8.0. A systematic knowledge of the organisms in the system is required for success. pH, ammonia, dissolved Oxygen, soluble Salts, alkalinity, nitrate are some of the measures for water quality which should be monitored from time to time.

Safe Materials

All the components used in the system should be made sure that they are safe for fish and humans:

- Polypropylene - labelled PP
- High Density Polyethylene - labelled HDPE
- High Impact ABS (Hydroponic Grow Trays)
- Stainless Steel barrels
- EPDM or PVC (poly vinyl chloride) pond liner (make sure it's UV resistant and avoid fire retardant material)
- Fibre glass tanks and grow beds
- Rigid white PVC pipe and fittings, black flexible PVC tubing, some ABS
- DO NOT use Copper – Its toxic to the fish

System Maintenance

- Fish should be fed daily and their health should be monitored regularly.
- Water quality should be tested (every other day for the first month, then about once a week, then as needed).
- Filter screens, filter tanks (if using), tubing, water pump, grow bed media, etc. should be cleaned out as and when needed.
- Plant health should be checked.
- Plants should be checked for bugs or nutrient deficiencies in a regular fashion.

System Start-up Checklist

- Type and size of system to build should be clearly decided
- Drawing to be done for designs, research where to get parts, plan
- Components should be brought and assembled properly
- Plants should be grown from seed or some source for seedlings should be found
- System should be filled with water and circulated (at least a week)
- About 20% of stocking density of fish should be added to the system
- Water quality should be monitored and partial water changes should be done as and when needed
- System should be maintained properly

Handy Tips and Tricks
- Gravel media should be washed before putting into the system – otherwise it will lead to very cloudy dirty water
- pH of the gravel media should be tested
- Vitamin C and an air pump to bubble out chlorine and chloramines from tap water should be used
- Worms (red wigglers) need to be used in media beds to breakdown solids and reduce anaerobic zones
- Cleaning products, pesticides, algaecides, fertilizers or like substances shouldn't be used in fish tanks or grow beds
- Plants should be sprayed with diluted vinegar and water solution if aphides infects the plants
- Direct sunlight on fish tanks should be avoided, the top should be covered to avoid algae and make fish happy
- More than 1/3 of water at a time shouldn't be changed. More than that will destroy the good bacteria in the system.
- Outdoor plants should be covered during a frost, and shade from the scorching summer sun. We need to make sure that we have backup power available for pumps and aerators

Benefits from Aquaponics
- Addresses issues on food safety
    - Produce do not contain the most common pathogen
- Maximizes the use of space
    - Diversified operations (fish and plants)
    - Ability to produce a large quantity of food in a small space
    - No land is needed
- Ease of operation
    - No weeding
    - No soil cultivation
    - Minimal watering
    - No pesticide application
    - Minimal maintenance and time spent

- Addresses issues on climate change
    - Conserves water.
    - No leaching of nutrients or waste to be pumped into the environment.
- A great educational tool to teach children grow food and care for living things.
- Operation is friendly to persons with physical disability.
- Products are higher in nutrients.

Future perspectives

Effortlessness design and management with nearly no energy and low equipment costs makes this system a motivating solution wherever land availability, flooding, productivity and ecological footprint are major issues. In addition the use of water weeds as a resource can undoubtedly increase livelihood opportunities in all those areas affected across the globe. Further research needs to address the nutrient dynamics of diverse growing media and to optimize system design and nutritional requirement of vegetables in those water bodies with limited dissolved nutrients. The possibilities of this integrated system are quite lofty and can provide sensitive benefits to small holders as well as big aquaculture enterprises. The potential of these systems is however not fully understood and interdisciplinary links and research can undeniably address many of the issues that are still buried.

Aquaponics is the amalgamation of aquaculture and hydroponic systems whereby nutrient rich waste water from the aquaculture system is engaged into the hydroponic system. The trends of new millennium in environmental regulation, are limiting amount of water which may be consumed or discharged. In aquaponics, wastewater from aquaculture is filtered and is recirculated into the system. Aquaponics presents an opening to tamper with the traditional fish farming, to bring in more money farm gate.

"The ultimate goal of farming is not the growing of crops, but the cultivation and perfection of human beings." — Masanobu Fukuoka, the One-Straw Revolution

REFERENCES

1. "Activities of NFDB". National Fisheries Development Board - Govt of India. 2008.
2. "Annual Report: India, 2008-2009". Department of Animal Husbandry Dairying and Fisheries, Ministry of Agriculture, Government of India. 2009.
3. "India - National Fishery Sector Overview". Food and Agriculture Organization of the United Nations. 2006.
4. "National Aquaculture Sector Overview: India". Food and Agriculture Organization of the United Nations. 2009
5. Anon. (2004 – 2005).:Annual Report, Indian Lac Agricultural Research.
6. Anon. (2005): State of Forest Report, Forest Survey of India, Govt. of India.
7. Bahl, K.N., 1943, *Pheretima*, The Indian Zoological Memories, Lucknow Publishing House, Lucknow, India.
8. Barik TK, Sahu B, Swain V (2008). Nanosilica-from medicine to pest control. *Parasitol. Res.* 103(2): 253-258.
9. Barnes, R.D., 1980, Invertebrate Zoology, W.B. Saunders Company, Philadelphia and London.
10. Barrington, E.J.W., 1974, Invertebrate structure and function, Thomas Nelson and Sons Ltd., London.
11. Bayer, F.M. and Owre, H.B., 1968, The Free Living Lower Invertebrates, The Macmillan Co., New York.
12. Berril, N.J.,1966, Biology in Action, Heinemann Educational Books Ltd., London, U.K.
13. Bhattacharyya A (2009). Nanoparticles-From Drug Delivery To Insect Pest Control. Akshar. 1(1): 1-7.
14. Bhattacharyya A., Bhaumik A., Pathipati U. R., Mandal S. and Epidi T. T. (2010). Nano-particles - A recent approach to insect pest Control, *African Journal ofBiotechnology.* 9(24), pp. 3489-3493.
15. Borradaile L.A., F.A. Potts and L.E.S. Estham, 1962, The Invertebrata, Asia Publishing House, Bombay, India.
16. Brown Jr., F.A., 1956, Selected Invertebrate Types, John Wiley and Sons, New York, U.S.A.
17. Buchsbaum, R., 1963, Animals without Backbones, Chicago University Press, Chicago, U.S.A.

18. Buffaloe, N.D., 1964, Principles of Biology, Prentice Hall of India Pvt. Ltd. New Delhi, India.

19. Bullough, W.S., 1958, Practical Invertebrate Anatomy, Macmillan and Company Ltd., London, U.K.

20. Carter, G.S., 1961, A General Zoology of the Invertebrates, Sidgwick and Jackson Ltd., London; U.K.

21. Chanu O. P. and Ibotombi N. (2011). Effects of 60Co gamma radiation on eggs of tasar silkworm, *Antheraeaproylei* (Lepidoptera), *Journal of Experimental Sciences, 2(9): 19-23*

22. Chattopadhyay S, and Pandey ON (2004):Lac Jungle creation : an approach for forest conservation. In "Lac industry-convergence for resurgence", ILRI, Ranchi.3-4.

23. Chattopadhyaya S; (2011): Introduction to lac and lac culture, *Annapurna press*, Department of forest biology and tree movement faculty of forestry birsa agriculture university, Ranchi.tech. bull.FBTI/01/2011.

24. Dales, R., Phillips, 1963, Annelids, Hutchinson University Library, London.

25. Das, S.K. (2011): Management Strategy for utilizing underutilized aquatic resources with Pro- Poor, Pro-Gender and Pro-Nature approach.(*In*:A.Sinha, S Dutta and B.K Mahapatra Eds) *Diversification of Aquaculture*, Narendra Publishing House, pp.355-363.

26. Das, S.K. and Padhi, S.N. (2010): Biological Considerations in Shrimp Farming. (*In*: L.R.Patro Ed.) *Aquatic Biodiversity*, Discovery Publishing House, NewDelhi. Pp.53-62. Felix, S. (2011): Vannamei (*Litopenaeus vannamei*) farming- A stich in time can avertanother aqua-calamity in India. *Fishing Chimes*, 30(12): 26-28.

27. Diver S, 2006. Aquaponics—Integration of Hydroponics with Aquaculture (Internet). ATTRA - National Sustainable Agriculture Information Service. Available from: <http://attra.ncat.org/attra-pub/PDF/aquaponic.pdf> (accessedon 02/4/2008)

28. Duan J., Li R., Cheng D., Fan W., Zha X., Cheng T., Wu Y., Wang J., Mita K., Xiang Z. and Xia Q. (2010). SilkDB v2.0: a platform for silkworm (*Bombyx mori*) genome biology, *Nucleic Acids Research*, 38, Database issue D453–D456.

29. Duong Tan Nhut , Nguyen Hoang Nguyen, Dang Thi Thu Thuy, "A novel in vitro hydroponic culture system for potato (Solanum tuberosum L.) microtuber production", ScienceDirect ,2006.

30. Edwards, C.A. and Lofty, J.R., Biology of Earthworm, Chapman and Hall, London.

31. Gardiner, M.S., 1972, The Biology of Invertebrates, McGraw Hill Book Company, New York.

32. Glover P.M (1937): Lac cultivation in India. ILRI, Namkum Ranchi. 119 pp.

33. Gordon A Chalmers, "Aquaponics and Food Safety", Lethbridge, Alberta April, 2004.
34. Groove, A.J.,G.E. Newel and J.D. Carthy, 1962, Animal Biology, University Tutorial Press Ltd., London, U.K.
35. Harmer, S.F., A.E. Shipley, 1959, Insects, Part II, The CNH, Vol. VI Macmillan and Company Ltd., London, U.K.
36. Hegner, R.W., and J.G. Engemann, 1968, Invertebrate zoology, Macmillan and Co. Ltd., New York, U.S.A.
37. Hickmann, C.P. 1961, Integrated Principles of Zoology, The C.V. Mosby Co., Louis.
38. Hickmann, C.P. 1973, Biology of The Invertebrates, The C.V. Mosby Co., Louis.
39. Hutchinson, W, Jeffrey, M, O'Sullivan, D., Casement, D. , Clarke, S., " Recirculating Aquaculture Systems: Minimum Standards for Design, Constructionand Management.", Inland Aquaculture Association of South Australia Inc. , 2004.
40. Hyman, L.H. 1967, The Invertebrates, Mollusca 1, Vol VI, McGraw Hill Book Company, New York.
41. IINRG Annual Report 2012-13, Indian Council of
42. Imms, A.D., Richards, O.W., and Davies, R.G., 1957, A general Text Book of Entomology, Methuen and Company Ltd., London, U.K.
43. Jennings, H.S., 1906, *Behaviour of Lower Organisms,* Coulmbia University Press, New York.
44. Jensen, M.H., "Hydroponics.", HortScience, 32(6):1018–1021. 1997.
45. Krishnaswami S (1960): Lac cultivation in India. Farm Bulletin, Directorate of Extension, Ministry of Food & Agriculture, Govt. of India, New Delhi, 36 pp.
46. Lakshmi H., Chandrashekharaiah, Ramesh Babu M., Raju P.J., Saha A.K.and Bajpai A.K. (2011). HTO5 x HTP5, The new bivoltine silkworm (*bombyx mori* l.) hybrid with thermo-tolerance for tropical areas, *InternationalJournalof Plant, Animal and Environmental Sciences,* 1(2):88-104.
47. Laverack, M.S., 1963, *The Physiology of Earthworms*, The Macmillan Company New York.
48. Little, V.A., 1967, *General and Applied Entomology*, Oxford and I.B.H. Publishing House, Calcutta, India.
49. Malcolm, J. , "What is aquaponics?", Backyard Aquaponics, Issue 1, Summer 2007.
50. Meglitsch, P.A., 1972, *Invertebrate Zoology*, Oxford University Press, New York, U.S.A.
51. Menon,N.R. (2007): Human Resource Development in fisheries in India and its relevance in the present context. *Proceedings of National Workshop on Fishfor All through Quality Fisheries Education,* College ofFisheries, Panagad, Kochi, Pp.3-17.

52. Morton, J.E., 1967, *Molluscs*, Hutchinson University Library, London.
53. Nagaraju J (2002). Application of genetic principles for improving silk production. *Curr. Sci.,* 83: 409-414.
54. Nichols M, "Aquaponics: Where One Plus One Equals Three", Massey University, Palmerston North, New zeeland, Maximum Yeld- Indoor gradening, UK January-February 2008
55. Norris D, Glover P.M. and Aldis R.W. (1934): Lac and the Indian Lac Research Institute, Namkum.53 pp.
56. Padhi, S. N. (2014). Application of biology for self employment. Edtd vol. Nanda Kishore Publication, Bhubaneswar.
57. Panda, Aparajita.; Panda, Sasmita and Das, S.K. (2016). Sustainable utilization of brackish water for shrimp and fish production. In Water For Survival. Nanda Kishore Publication. Pp 18-22. ISBN-978-81-932445-0-1.
58. Panda, Sasmita (2016). A review on induced breeding in fishes. Ijbio. Vol. 5, (5). Pp 4579-4588. ISSN-2278-778X.
59. Panda, Sasmita (2016). Composite fish culture for gainful employment. Ijbio. Vol. 5, (6). Pp 4593-4596. ISSN-2278-778X.
60. Panda, Sasmita.; Panigrahi, G.K. and Padhi, S.N. (2015): Limnological characteristics of ponds for improvement of livelihood of fishermen through aquaculture. Ijbio. Vol. 4, (12). Pp 4590-4594. ISSN-2278-778X.
61. Pantanella, E., "Pond aquaponics: new pathways to sustainable integrated aquaculture and agriculture", Aquaculture News, May 2008
62. Parker, T.J. and William, A. Haswell Edited by A.J., Marshall And W.D. Williams. (7th edition), 1972, *A Text Book of Zoology: Invertebrates*, English Language Book Society and Macmillan Company, London.
63. Patton, R.L., 1963, *Introductory Insect Physiology*, W.B. Saunders Co., Philadelphia.
64. Pon, R.D, 1968, *The Biology of Mollusca*, Pergamon Press, New York.
65. Ponniah, A.G. (2010): Shrimp Culture. *Hand book of Fisheries and Aquaculture*, Indian Council of Agricultural Research, New Delhi, pp-348-360.
66. Ponnuvel KM, Nakazawa H, Furukawa S, Asaoka A, IshibashiJ, TaakaH, Yamakawa M (2003). A Lipase isolated from the silkworm shows antiviral activity against NPV. *J. Virol.* 77(19): 10725-10729.
67. Rakocy, James E.; Masser, Michael P.; Losordo, Thomas M. (November 2006). Recirculating aquaculture tank production systems: Aquaponics — integrating fish and plant culture (454). Southern Regional Aquaculture Center.

68. Ravichanran, P. (2006): Shrimp Farming. *Hand book of Fisheries and Aquaculture*, Indian Council of Agricultural Research, New Delhi, pp-392-403. Research Institute, Ranchi.

69. Russel-Hunter, W.D., 1968, *A Biology of Higher Invertebrates*, The Macmillan Co., New York.

70. Russel-Hunter, W.D., 1968, *A Biology of Lower Invertebrates*, The Macmillan Co., New York.

71. Sen P., Maurya R.C. and Gokulpure R.C. (1981):On some hosts of lac insects, *Kerria lacca* (Kerr). Indian Forester. 107 (9) : 583- 584.

72. Sing R. ;(2006): Lac culture, Applied Zoology.

73. Snodgrass, R.E., 1952, *A Text Book of Arthropod Anatomy*, Comstock Publishing Associates, Ithaca, New York, U.S.A.

74. Soulsby, E.J.L., 1976, *Helminths, Arthropds and Protozoa of Domesticated Animals*, The English Language Book Socirty, London.

75. Stiles, K.A., R.W. Henger, and R.A. Boolootian, 1969, *College Zoology*, American Publishing Co., Pvt. Ltd., New Delhi, Bombay, Calcutta, India.

76. Storer, T.L., and R.L. Usinger, 1965, *General Zoology*, McGraw Hill Book Company, New York, London.

77. Thangavelu K, Sinha R, Mohan RK (2003). Silkworm Germplasm and their Potential Use. Proceeding of Mulberry Silkworm Breeders Summit; Hindupur, India. pp. 14-23.

78. Vasudha BC, Aparna HS, Manjunatha HB. (2006). Impact of heat shock on heat shock proteins expression, biological and commercial traits of *Bombyxmori. Insect Science* 13: 243-250.

79. Verma, P.S., 2007, *A Manual of Practical Zoology Invertebrates,* S. Chand & Co. Ltd., New Delhi, India.

80. Wigglesworth, V.B., 1965, *The Principles of Insect Physiology,* English Language Book Society and Methuen and Company Ltd., London, U.K.

81. Wilmoth, Jmaes H., 1967, *Biology of Invertebrates,* Prentice Hall, Inc. Englewood Cliffs, New Jersey.

Web Resources for figure

1. aquaponicsgardening.ning.com – Community blog
2. attra.ncat.org/attra-pub/aquaponic.html - ATTRA
3. https://en.wikipedia.org/wiki.
4. https://seaworld.org/en/animal-info books.
5. sweetwater-organic.com/blog - Milwaukee, Wi

Wikipedia

6. wwf.panda.org.
7. www.aquaponics.com – Nelson and Pade, Montello, Wi
8. www.aquaponics.net.au – Murray Hallam, Australia
9. www.aquaponicsusa.com - California
10. www.backyardaquaponics.com – Joel Malcolm, Australia
11. www.coloradoaquaponics.com
12. www.friendlyaquaponics.com – Hawaii
13. www.growingpower.org – Will Allen, Milwaukee, Wi
14. www.uvi.edu – University of Virgin Island

Aquaponic Resources

1. Aquasafra, Inc.(http://www.tilapiaseed.com/)
2. Aquatic Eco-systems, Inc.(http://www.aquaticeco.com/)
3. Blackwater Creek Koi Farms (http://www.koisale.com/)
4. Harbor Branch Oceanographic Institute (http://www.fau.edu/hboi/)